AF895922

Unramified Brauer Group and Its Applications

Translations of
MATHEMATICAL
MONOGRAPHS

Volume 246

Unramified Brauer Group and Its Applications

Sergey Gorchinskiy
Constantin Shramov

AMERICAN MATHEMATICAL SOCIETY
Providence, Rhode Island

EDITORIAL COMMITTEE

Lev Birbrair Pavel Etingof (Chair)
Toshiyuki Kobayashi Shou-Wu Zhang

2010 *Mathematics Subject Classification.* Primary 16K50, 14E08; Secondary 14M20, 14G05, 20J06, 12G05.

For additional information and updates on this book, visit
www.ams.org/bookpages/mmono-246

Library of Congress Cataloging-in-Publication Data

Names: Gorchinskiy, Sergey, 1982- author. | Shramov, Constantin, author.
Title: Unramified Brauer group and its applications / Sergey Gorchinskiy, Constantin Shramov, authors.
Description: Providence, Rhode Island : American Mathematical Society, [2018] | Series: Translations of mathematical monographs ; volume 246 | Includes bibliographical references and index.
Identifiers: LCCN 2018005037 | ISBN 9781470440725 (alk. paper)
Subjects: LCSH: Brauer groups. | Associative algebras. | AMS: Associative rings and algebras – Division rings and semisimple Artin rings – Brauer groups. msc | Algebraic geometry – Birational geometry – Rationality questions. msc | Algebraic geometry – Special varieties – Rational and unirational varieties. msc | Algebraic geometry – Arithmetic problems. Diophantine geometry – Rational points. msc | Group theory and generalizations – Connections with homological algebra and category theory – Cohomology of groups. msc | Field theory and polynomials – Homological methods (field theory) – Galois cohomology. msc
Classification: LCC QA251.5 .G67 2018 | DDC 512/.46–dc23
LC record available at https://lccn.loc.gov/2018005037

DOI: http://dx.doi.org/10.1090/mmono/246

Copying and reprinting. Individual readers of this publication, and nonprofit libraries acting for them, are permitted to make fair use of the material, such as to copy select pages for use in teaching or research. Permission is granted to quote brief passages from this publication in reviews, provided the customary acknowledgment of the source is given.

Republication, systematic copying, or multiple reproduction of any material in this publication is permitted only under license from the American Mathematical Society. Requests for permission to reuse portions of AMS publication content are handled by the Copyright Clearance Center. For more information, please visit www.ams.org/publications/pubpermissions.

Send requests for translation rights and licensed reprints to reprint-permission@ams.org.

© 2018 by the authors. All rights reserved.
The American Mathematical Society retains all rights
except those granted to the United States Government.
Printed in the United States of America.

∞ The paper used in this book is acid-free and falls within the guidelines
established to ensure permanence and durability.
Visit the AMS home page at http://www.ams.org/

10 9 8 7 6 5 4 3 2 1 23 22 21 20 19 18

To Alexey Nikolaevich Parshin on his 75th birthday, with respect and gratitude

Contents

Preface

This book is an extended version of the notes of a reading seminar "Arithmetic methods in algebraic geometry" run by the authors at the Steklov Mathematical Institute in Spring 2011. The goal of the book is to give an introduction to the theory of unramified Brauer groups and their applications to stable rationality, starting with the most basic concepts of group cohomology. For this reason, we omit many popular topics that are already well covered in textbooks (Galois cohomology, Brauer groups, etc). Instead we give more attention to applications of unramified Brauer groups to stable non-rationality, and to an example of a non-rational stably rational variety. As far as we know, these topics are not covered in detail in standard textbooks, and the proofs presented in the original sources require substantial effort to understand.

The style of our seminar suggested delivering the material through problems and exercises. We have tried to split the proofs of all facts that we need into relatively simple steps and provide detailed hints for all non-trivial points. This gives us hope that studying our book will be no more difficult (or at least not much more difficult) than reading a usual textbook, not to mention research articles. Most of the book is accessible to those who are familiar with basic algebra, Galois theory, and fundamental notions of algebraic geometry.

In Chapter 1 we collect the necessary definitions and facts concerning cohomology of abstract groups. The same is done in Chapter 2 for Galois cohomology and in Chapter 3 for Brauer groups. Since the significance of these topics is much broader than their applications to stable rationality and they constitute an important part of modern mathematical culture, we recommend that the reader continue their study with the help of canonical sources. For group cohomology we recommend Chapter IV in the book [**CF67**], for Galois cohomology the book [**Ser65**] and Chapter V in [**CF67**], and for Brauer groups Chapter X of the book [**Ser79**] and [**Bou58**]. In Chapter 4 we focus on the Brauer group of a discrete valuation field and in particular define the unramified Brauer group. For further reading on these topics, we refer to the book [**Ser79**] and §1 of Chapter VI in the book [**CF67**]. Besides these references, most of the material of Chapters 1–4 is covered in much more detail in the textbook [**GS06**]. The interested reader can also find an accessible account of Galois cohomology in [**Ber10**].

In Chapter 5 we present the example of a quotient variety $X = V/G$, where G is a finite group and V is a representation of G over an algebraically closed field k of characteristic zero, which can be proved to be non-rational (and even not stably rational) using the notions introduced earlier. The obstruction we use is non-triviality of the unramified Brauer group of the field $k(X)$, that is, of the invariant field $k(V)^G$. Examples of this kind first appeared in the works of

D. Saltman [**Sal84**] and F. A. Bogomolov [**Bog87**], but we adopt the simpler approach taken by I. R. Shafarevich in [**Sha90**]. The variety X has relatively large dimension; one may be interested in whether there are similar examples in lower dimensions. It turns out that this is possible already for some threefolds, based on a completely different construction from the one given in Chapter 5. We present such an example in Chapter 7: the well-known construction of a non-rational singular double cover of $\mathbb{P}^3$ branched over a quartic. This variety was first described in the paper [**AM72**] by M. Artin and D. Mumford, but we take a more algebraic approach due to M. Gross (see [**AM96**, Appendix]). Before presenting the construction, we introduce the Clifford invariant and spend some time on auxiliary results about quadrics over non-algebraically closed fields in Chapter 6. More on quadrics over non-algebraically closed fields can be found in the book [**EKM08**]. In Chapter 7 we also provide a unirationality construction for a double cover of $\mathbb{P}^3$ branched over a quartic (here we mostly follow the proof of Theorem IV.7.7 in the book [**Man86**]). A detailed survey of stable (non-)rationality results for quotient varieties similar to the ones considered in Chapter 5 is contained in [**CTS07**]; one can also find references to many original works on the topic therein. We also recommend that the reader have a look at the short survey [**BT17**]. For a discussion of results on obstructions to stable rationality appearing from the Artin–Mumford construction, we refer the reader to the survey [**Pir16**].

In Chapter 8 we introduce Weil restriction and discuss its main properties, and we also establish some properties of algebraic tori that will be used in Chapter 9. More details on algebraic tori are available in the book [**Vos98**]. Chapter 8 also discusses the notion of universal torsor and some basic properties of Châtelet surfaces. Since we already know examples of varieties which are not stably rational, it is natural to ask whether or not stable rationality is actually the same as rationality. It turns out that it is not the same, but producing an example that separates these two concepts is not easy at all. This is done in Chapter 9, following the paper [**BCTSSD85**] by A. Beauville, J.-L. Colliot-Thélène, J.-J. Sansuc, and P. Swinnerton-Dyer. At the end of Chapter 9 we provide an argument of N. Shepherd-Barron from [**SB04**] that slightly enhances the construction of [**BCTSSD85**]. Throughout Chapter 9 we try to use geometric language and to avoid coordinates and explicit equations as far as possible, which we hope will make our exposition a bit more transparent than that of [**BCTSSD85**] and [**SB04**].

Chapters 10 and 11 are devoted to one more application of unramified Brauer groups, namely, to Brauer–Manin obstructions. The main purpose of Chapter 10 is to provide some motivation for this: we discuss a proof of the classic Minkowski–Hasse theorem for quadrics (to be more precise, we deduce this theorem from the fundamental facts of class field theory). Our exposition mostly follows Chapter IV of the book [**Ser70**], but we try to use more geometric language when reducing the multi-dimensional case to the one-dimensional case. In Chapter 11 we define the Brauer–Manin obstruction and use it to produce a counterexample to an analog of the Minkowski–Hasse theorem for curves of genus 1. More on the Brauer–Manin obstruction can be found in the surveys [**Sko01**], [**Poo17**], and [**Wit16**] (see also the brief exposition in [**MP05**, 5.2.3]).

Appendix A contains a collection of references to the main results on étale cohomology that are necessary for interpretation of Brauer groups in étale terms (see [**Dan96**] or [**Mil80**] for more details on these results). Those who have a taste for exploring primary sources may wish to take a look at the text [**Gro95b**] by A. Grothendieck, where this very approach was used to introduce the unramified Brauer group for the first time.

As one might expect, we were not able to pay enough attention to many topics related to unramified Brauer groups (in particular, to the study of stable rationality, which became remarkably active in recent years). To (partially) fill this gap, we conclude most of the chapters of the book with lists of references for an interested reader, sometimes with brief explanations about their connections to the material covered in the chapter. In several cases (especially in Chapters 5, 7, and 11) we also tried to include references to recent works, because the techniques mentioned in these chapters are still being developed and actively applied.

We would like to thank all the participants of our seminar for fruitful interactions and a stimulating atmosphere. While preparing these notes, we benefited from advice and discussions with J.-L. Colliot-Thélène, A. Fonarev, N. Howell, A. Kuznetsov, I. Marshall, I. Netay, Yu. Prokhorov, S. Rybakov, T. Shabalin, E. Shinder, A. Skorobogatov, M. Temkin, D. Testa, A. Trepalin, A. Vishik, V. Vologodsky, and V. Zhgun.

Our work was partially supported by the Dynasty foundation of D. Zimin, by the Russian Academic Excellence Project "5-100", and by the Laboratory of Mirror Symmetry NRU HSE, RF government grant, ag. № 14.641.31.0001.

Sergey Gorchinskiy and Constantin Shramov

Notation

$\mathbb{Z}$ — the ring of integers

$\mathbb{Q}$ — the field of rational numbers

$\mathbb{R}$ — the field of real numbers

$\mathbb{C}$ — the field of complex numbers

$\mathbb{F}_q$ — the finite field of q elements

$\mathbb{Z}_p$ — the ring of p-adic integers

$\mathbb{Q}_p$ — the field of rational p-adic numbers

$\widehat{\mathbb{Z}}$ — the profinite completion of the (additive) group $\mathbb{Z}$

$k(t_1, \ldots, t_n)$ — the field of rational functions in independent variables $t_1, \ldots, t_n$ over a field k

$k((t))$ — the field of Laurent series in a variable t over a field k

$\mathrm{Hom}(X, Y)$ — the set of morphisms from X to Y (in a category that is usually obvious from the context)

$\mathrm{Aut}(X)$ — the automorphism group of X

S_n — the symmetric group on n letters

$\mathrm{Stab}_G(x)$ — the stabilizer in a group G of an element x of some set X with an action of G

G/H — the set of left cosets in a group G of its subgroup H

A_n — the n-torsion of an abelian group A for a positive integer n

$\mathbb{Z}[S]$ — the free abelian group generated by a set S

$\langle S \rangle$ — the subgroup generated by a subset S of some group, or a two-sided ideal generated by a subset S of some associative algebra

M^G — the group of G-invariant elements in a G-module M

$\mathrm{Hom}_G(M, M')$ — the group of morphisms between G-modules M and M'

$H^i(G, M)$ — the ith cohomology group of a group G with coefficients in a G-module M

Z/Γ — the set of orbits of a group Γ acting on a set Z

$\mathrm{char}(R)$ — the characteristic of a ring R

R^* — the multiplicative group of invertible elements of a ring R

$K_2(K)$ — the second Milnor K-group of a field K

$\bar{K}$ — the algebraic closure of a field K

K^{sep} — the separable closure of a field K

μ_n — the group of nth roots of unity in K^{sep}, where n is coprime to $\mathrm{char}(K)$

$\dim_K(V)$ — the dimension of a vector space V over a field K

$[L : K] = \dim_K(L)$ — the degree of a finite extension of fields $K \subset L$

$\mathrm{Nm}_{L/K} \colon L^* \to K^*$ — the Galois norm for a separable finite extension of fields $K \subset L$

$\mathrm{Gal}(L/K)$ — the Galois group of a Galois extension $K \subset L$

$G_K = \mathrm{Gal}(K^{sep}/K)$ — the absolute Galois group of a field K

GL_n — the group of invertible $n \times n$ matrices

$\mathbb{G}_m = \mathrm{GL}_1$ — the multiplicative group scheme

SL_n — the group of invertible $n \times n$ matrices with trivial determinant

$\mathrm{PGL}_n = \mathrm{GL}_n / \mathbb{G}_m$ — the group of invertible $n \times n$ matrices modulo scalar matrices, that is, the automorphism group of an $(n-1)$-dimensional projective space

$\mathbb{P}^n$ — the projective space of dimension n

$\mathcal{O}_{\mathbb{P}^n}(r) = \mathcal{O}_{\mathbb{P}^n}(1)^{\otimes r}$ — the rth tensor power of the line bundle $\mathcal{O}_{\mathbb{P}^n}(1)$ on $\mathbb{P}^n$ that is dual to the tautological bundle $\mathcal{O}_{\mathbb{P}^n}(-1)$, where r is a positive integer

$V_L = L \otimes_K V$ — the scalar extension of a vector space V over a field K, where $K \subset L$ is some field extension

$\mathrm{GL}(V)$ — the group of K-linear automorphisms of a vector space V over K

$\mathbb{P}(V) \cong \mathbb{P}^{n-1}$ — the projectivization of an n-dimensional vector space V over a field K (the set of K-points of $\mathbb{P}(V)$ is the set of one-dimensional subspaces of V)

$\mathrm{Spec}(R)$ — the spectrum of a commutative unital ring R

$k[X]$ — the ring of regular functions on a variety X over a field k

$k(X)$ — the field of rational functions on an irreducible variety X over a field k

$\dim(X)$, $\dim_k(X)$ — the dimension of a variety X over a field k

$\mathrm{Div}(X)$ — the group of divisors on a smooth variety X

$\mathrm{Pic}(X)$ — the Picard group of a variety X

$\mathbb{T}_x(X) \subset \mathbb{P}^n$ — the embedded projective tangent space at a point x to a projective variety $X \subset \mathbb{P}^n$

X_K — the scalar extension of a variety X defined over a field k, where $k \subset K$ is some field extension

$X(K)$ — the set of K-points of a variety X defined over a field k, where $k \subset K$ is some field extension

$\mathrm{M}_n(A)$ — the algebra of $n \times n$ matrices with entries in an associative algebra A

$\mathrm{Br}(K)$ — the Brauer group of a field K

$\mathrm{Br}(L/K) = \mathrm{Ker}\big(\mathrm{Br}(K) \to \mathrm{Br}(L)\big)$ — the relative Brauer group of a field extension $K \subset L$

$b(X) \in \mathrm{Br}(K)$ — the class of a Severi–Brauer variety X defined over a field K

$\mathrm{res}\colon \mathrm{Br}(K) \to \mathrm{Hom}(G_\kappa, \mathbb{Q}/\mathbb{Z})$ — the residue map for a complete discrete valuation field K with a perfect residue field κ

$\mathcal{O}_K$ — the valuation ring in a discrete valuation field K, or the ring of integers in a number field K

K_v — the completion of a field K with respect to a discrete (or, more generally, multiplicative) valuation v

$\mathrm{res}_v\colon \mathrm{Br}(K) \to \mathrm{Hom}(G_{\kappa_v}, \mathbb{Q}/\mathbb{Z})$ — the residue corresponding to a discrete valuation v of a field K, where κ_v is a residue field of K_v and is assumed to be perfect

$\mathrm{res}_D\colon \mathrm{Br}\big(k(X)\big) \to \mathrm{Hom}(G_{k(D)}, \mathbb{Q}/\mathbb{Z})$ — the residue corresponding to a discrete valuation given by a reduced irreducible divisor D on an irreducible (normal) variety X over a field k of characteristic zero

$\mathrm{Br}^{nr}(K)$ — the unramified Brauer group of a finitely generated field K over a field k of characteristic zero

$\mathrm{Br}^{nr}(X)$ — the unramified Brauer group of an irreducible (normal) variety X over a field k of characteristic zero

$\mathrm{Br}(X)$ — the Brauer group of a variety or a scheme X

$\mathrm{d}(Q) \in K^*/(K^*)^2$ — the discriminant of an even-dimensional quadric Q over a field K, where the characteristic of K is different from 2

$\mathrm{cl}(Q) \in \mathrm{Br}(K)$ — the Clifford invariant of a quadric Q over K with trivial $\mathrm{d}(Q)$, where the characteristic of K is different from 2

$R_{K/k}(Y)$ — the Weil restriction for a variety Y over K and a separable finite extension of fields $k \subset K$

$K^* = R_{K/k}(\mathbb{G}_m)$ — the algebraic torus over a field k for a separable finite extension of fields $k \subset K$, whose set of k-points is $K \setminus \{0\}$

$T^\vee$ — the dual G_k-module of an algebraic torus T over a field k

$M^\vee$ — the dual algebraic torus over k to a G_k-module M, which is free and finitely generated as an abelian group

Part I

Preliminaries on Galois cohomology

CHAPTER 1

Group Cohomology

1.1. Definition and basic properties

Given two sets X and Y, we denote by $\mathrm{Map}(X,Y)$ the set of all maps from X to Y. If G and H are groups, we denote by $\mathrm{Hom}(G,H)$ the set of all homomorphisms from G to H. If A is an abelian group, then the set $\mathrm{Hom}(G,A)$ is an abelian group as well.

For an arbitrary group G, we denote by e the neutral element in G and by Mod_G the category of (left) *G-modules*, that is, abelian groups M with an action of G commuting with addition in M. The image of an element m of a G-module M under the action of $g \in G$ is denoted by ${}^g m$. In particular, one has

$${}^g({}^h m) = {}^{gh} m\,.$$

By $\mathrm{Hom}_G(M,M')$ we denote an abelian group of morphisms in the category Mod_G from a G-module M to a G-module M'. By M^G we denote a subgroup in M that consists of all G-invariant elements. We say that a G-module M is *trivial* if G acts as the identity on M. The group $\mathbb{Z}$ is usually assumed to have the structure of a trivial G-module.

For an arbitrary set S we denote by $\mathbb{Z}[S]$ the lattice with basis labelled by elements of S. In particular, if there is a G-action on S, then $\mathbb{Z}[S]$ becomes a G-module. Also, $\mathbb{Z}[G]$ is a ring, and we call it the *group ring* of the group G; the natural action of G on itself by left translations defines a G-module structure on $\mathbb{Z}[G]$.

EXERCISE 1.1.1. **Category of G-modules**

(i) Show that Mod_G is an abelian category and that Mod_G is equivalent to the category of left modules over the group ring $\mathbb{Z}[G]$.

(ii) Show that there is a canonical isomorphism

$$\mathrm{Hom}_G(\mathbb{Z},M) \cong M^G$$

and that taking G-invariants is a left-exact functor.

Throughout the book we will extensively use cohomology groups $H^i(G,M)$ of a group G with coefficients in the module M. Definition 1.1.2 below relies on the notion of the functor Ext in an abelian category. A reader unfamiliar with this concept can consider Exercise 1.1.3(iv) and (v) as the definition of cohomology groups $H^i(G,M)$.

DEFINITION 1.1.2. *Cohomology groups of a group G with coefficients in a G-module M* are given by the formula

$$H^i(G,M) = \mathrm{Ext}^i_G(\mathbb{Z},M), \quad i \geqslant 0\,,$$

where the functor Ext is taken in the abelian category of G-modules.

Therefore, $H^i(G, M)$ are right derived functors of the functor of G-invariants. In particular, a morphism

$$f\colon M \to M'$$

of G-modules gives rise to homomorphisms of cohomology groups

$$f\colon H^i(G, M) \to H^i(G, M'),$$

and a short exact sequence of G-modules

$$0 \to M' \to M \to M'' \to 0$$

gives rise to a long exact sequence

$$0 \to H^0(G, M') \to H^0(G, M) \to H^0(G, M'') \\ \to H^1(G, M') \to H^1(G, M) \to H^1(G, M'') \to \ldots$$

of abelian groups.

EXERCISE 1.1.3. **Standard complex**

(i) Prove that the G-module

$$\mathbb{Z}[\underbrace{G \times \ldots \times G}_{i}] = \{\textstyle\sum n_{g_1 \ldots g_i} \cdot (g_1, \ldots, g_i) \mid n_{g_1 \ldots g_i} \in \mathbb{Z}\}$$

with coordinatewise multiplication by elements of G on the left is a projective (and even free) object in Mod_G for every $i \geqslant 1$.

Hint. Consider the elements $(e, g_2, \ldots, g_i)$ in $\mathbb{Z}[G^{\times i}]$.

(ii) Define linear maps

$$\partial\colon \mathbb{Z}[G^{\times(i+1)}] \to \mathbb{Z}[G^{\times i}], \quad i \geqslant 0,$$

by formulas

$$\partial(g_1, \ldots, g_{i+1}) = \sum_{j=1}^{i+1} (-1)^{j+1} (g_1, \ldots, \hat{g}_j, \ldots, g_{i+1}),$$

where $\hat{g}$ means we omit the element g; in particular, for $i = 0$ this gives the map

$$\mathbb{Z}[G] \to \mathbb{Z}, \quad \sum n_g g \mapsto \sum n_g.$$

Prove that ∂ is an isomorphism of G-modules and that $\partial \circ \partial = 0$. This defines a complex $\mathbb{Z}[G^\bullet]$ of G-modules (here we write $G^\bullet$ instead of $G^{\times\bullet}$ for brevity).

(iii) Prove that the complex $\mathbb{Z}[G^\bullet]$ is exact, so that it is a projective resolution of the trivial G-module $\mathbb{Z}$.

Hint. Use the maps

$$G^i \to G^{i+1}, \quad (g_1, \ldots, g_i) \mapsto (e, g_1, \ldots, g_i)$$

to construct a contracting homotopy.

(iv) Given a G-module M, consider the complex $K^\bullet$ with terms

$$K^i = \mathrm{Hom}_G\big(\mathbb{Z}[G^{\times(i+1)}], M\big), \quad i \geqslant 0.$$

Show that

$$H^i(G, M) \cong H^i(K^\bullet)$$

for every $i \geqslant 0$.

(v) Compute explicitly the terms of the complex $K^\bullet$ from part (iv) and use this to prove that the group $H^i(G,M)$ is canonically isomorphic to the ith cohomology group of the complex

$$0 \to M \xrightarrow{d} \mathrm{Map}(G,M) \xrightarrow{d} \ldots \xrightarrow{d} \mathrm{Map}(G^{\times i},M) \xrightarrow{d} \ldots, \tag{1.1}$$

where the differential d is given by the formula

$$(d\varphi)(g_1,\ldots,g_{i+1}) = {}^{g_1}\big(\varphi(g_2,\ldots,g_{i+1})\big) + \sum_{j=1}^{i}(-1)^j\varphi(g_1,\ldots,g_jg_{j+1},\ldots,g_{i+1}) + (-1)^{i+1}\varphi(g_1,\ldots,g_i)\,.$$

In particular, in degree zero one has

$$(dm)(g) = {}^g m - m, \qquad m \in M\,.$$

EXERCISE 1.1.4. **Cohomology groups H^0, H^1, and H^2**

(o) Check that $H^0(G,M) \cong M^G$.

(i) Show that 1-cocycles of the complex (1.1) are *crossed homomorphisms* from G to M, that is, maps $\varphi\colon G \to M$ that satisfy the condition

$$\varphi(gh) = \varphi(g) + {}^g\big(\varphi(h)\big)$$

for all $g,h \in G$. Furthermore, a 1-cocycle φ is a coboundary if and only if there exists an element $m \in M$ such that

$$\varphi(g) = {}^g m - m$$

for all $g \in G$. In particular, the first cohomology group $H^1(G,M)$ of a trivial G-module M is isomorphic to the group of homomorphisms $\mathrm{Hom}(G,M)$.

(ii) Prove that the group $H^2(G,M)$ is bijective with the set of isomorphism classes of extensions

$$0 \to M \to \widetilde{G} \xrightarrow{\pi} G \to 0\,,$$

where an isomorphism between two extensions is defined as an isomorphism between the middle terms of the corresponding exact sequences which induces identity maps on the outer terms.

Hint. An extension as above gives rise to a 2-cocycle

$$\varphi(g,h) = \tilde{g}\tilde{h}(\widetilde{gh})^{-1}\,,$$

where $g \mapsto \tilde{g}$ is an arbitrary set-theoretic section of the homomorphism π. The structure of a G-module on M is given by conjugation, that is,

$${}^g m = \tilde{g}m\tilde{g}^{-1}, \qquad m \in M, g \in G\,.$$

In the opposite direction, given a cocycle φ one defines the group $\widetilde{G}$ as the set $M \times G$ with the group operation

$$(m,g)\cdot(n,h) = \big(m + {}^g n + \varphi(g,h), gh\big), \qquad m,n \in M,\ g,h \in G\,.$$

The cocycle condition is equivalent to associativity of this operation.

(iii) Consider a morphism $f\colon M \to M'$ of G-modules and an element $\alpha \in H^2(G, M)$ with corresponding extension

$$0 \to M \xrightarrow{\iota} \widetilde{G} \xrightarrow{\pi} G \to 0 .$$

Note that the homomorphism π gives a $\widetilde{G}$-module structure on M', and show that the image of the embedding

$$(f, \iota)\colon M \to M' \rtimes \widetilde{G}$$

is a normal subgroup. Check that the element

$$f(\alpha) \in H^2(G, M')$$

corresponds to an extension

$$0 \to M' \to \widetilde{G}' \to G \to 0 ,$$

where

$$\widetilde{G}' = \big(M' \rtimes \widetilde{G}\big)/(f, \iota)(M) .$$

(iv) Show that for a prime number p, one has an isomorphism

$$H^2(\mathbb{Z}/p\mathbb{Z}, \mathbb{Z}/p\mathbb{Z}) \cong \mathbb{Z}/p\mathbb{Z} .$$

Hint. Use the classification of groups of order p^2. Actually, the same holds if one replaces p by an arbitrary integer; cf. Exercise 1.1.5(i) below.

EXERCISE 1.1.5. **Cohomology of cyclic groups**

(i) Let G be a finite cyclic group of order n generated by an element s. Prove that the complex

$$\ldots \to \mathbb{Z}[G] \to \ldots \to \mathbb{Z}[G] \xrightarrow{s-1} \mathbb{Z}[G] \xrightarrow{\mathrm{N}} \mathbb{Z}[G] \xrightarrow{s-1} \mathbb{Z}[G] \to \mathbb{Z} \to 0$$

is a projective resolution of $\mathbb{Z}$, where N denotes the *norm map*, that is,

$$\mathrm{N} = \sum_{g \in G} g \in \mathbb{Z}[G] .$$

Applying the functor $\mathrm{Hom}_G(-, M)$, deduce that the group $H^i(G, M)$ is isomorphic to the ith cohomology group of the complex

$$0 \to M \xrightarrow{s-1} M \xrightarrow{\mathrm{N}} M \xrightarrow{s-1} \ldots \to M \to \ldots .$$

Thus we have the following isomorphisms:

$$H^i(G, M) \cong M^G/\mathrm{N}(M)$$

for even $i > 0$, and

$$H^i(G, M) \cong \mathrm{Ker}(\mathrm{N})/(s-1)M$$

for odd i. In particular, this formula for $H^i(G, M)$ gives another approach to Exercise 1.1.4(iv). Check that these isomorphisms are functorial with respect to M.

(ii) Let $M = \mathbb{Z}$ be a trivial G-module. Show that under the isomorphism

$$H^2(G, M) \cong M^G/\mathrm{N}(M),$$

the class of the element $1 \in \mathbb{Z}$ in the quotient group

$$M^G/\mathrm{N}(M) \cong \mathbb{Z}/n\mathbb{Z}$$

corresponds to the extension

$$0 \longrightarrow \mathbb{Z} \xrightarrow{n} \mathbb{Z} \xrightarrow{\lambda} G \to 0\,,$$

where $\lambda(1) = s$. In particular, this extension depends on the choice of a generator s. Similarly, the isomorphisms from part (i) depend on s. (When does one use the generator in the construction of these isomorphisms?)

(iii) Let M be an arbitrary G-module, and let $m \in M^G$ be a G-invariant element. Prove that the class of m in the quotient group

$$M^G/\mathrm{N}(M) \cong H^2(G, M)$$

corresponds via part (i) to the extension

$$0 \to M \to \widetilde{G} \xrightarrow{\pi} \mathbb{Z}/n\mathbb{Z} \to 0\,,$$

where

$$\widetilde{G} = \big(M \rtimes \mathbb{Z}\big)/\langle(m, n)\rangle$$

and the action of $\mathbb{Z}$ on M is via the homomorphism λ from part (ii).

Hint. Start with the case of a trivial module $M = \mathbb{Z}$ and an element $m = 1 \in \mathbb{Z}$, and use part (ii). For arbitrary M and m, consider the morphism

$$\mathbb{Z} \to M, \quad l \mapsto l \cdot m$$

of G-modules, and use Exercise 1.1.4(iii).

(iv) Assuming the notation of part (iii), show that there is a group isomorphism

$$\widetilde{G} \cong M \times \{1, \sigma, \ldots, \sigma^{n-1}\}\,,$$

where the group law on the right-hand side is defined by relations $\sigma^n = m$ and

$$\sigma \cdot m' = {}^{s}(m') \cdot \sigma, \quad m' \in M\,.$$

The homomorphism

$$\pi\colon \widetilde{G} \to \mathbb{Z}/n\mathbb{Z} \cong G$$

from part (iii) sends σ to s^{-1}.

EXERCISE 1.1.6. **Coboundary maps**
Consider a short exact sequence

$$0 \to M' \to M \xrightarrow{f} M'' \to 0$$

of G-modules and the corresponding long exact sequence of cohomology groups. Denote by

$$\delta\colon H^i(G, M'') \to H^{i+1}(G, M')$$

the coboundary maps in this long exact sequence.

(i) Show that for every element

$$m'' \in (M'')^G \cong H^0(G, M'')$$

one has

$$(\delta m'')(g) = -m + {}^g m\,,$$

where $m \in M$ is an arbitrary preimage of m''.

(ii) Show that for a 1-cocycle $\varphi\colon G \to M''$ one has

$$(\delta\varphi)(g,h) = \tilde{\varphi}(g) + {}^g\big(\tilde{\varphi}(h)\big) - \tilde{\varphi}(gh)\,,$$

where $\tilde{\varphi}(g) \in M$ is an arbitrary preimage of an element $\varphi(g)$, $g \in G$.

(iii) Suppose that M', M, and M'' are trivial G-modules. Consider a homomorphism $\varphi\colon G \to M''$. Show that the element $\delta\varphi \in H^2(G, M')$ corresponds to the extension (see Exercise 1.1.4)

$$0 \to M' \to \widetilde{G} \to G \to 1\,,$$

where

$$\widetilde{G} = M \times_{M''} G = \{(m, g) \in M \times G \mid f(m) = \varphi(g)\}\,.$$

(iv) Suppose that G is a finite cyclic group of order n generated by an element s. Show that under the identifications from Exercise 1.1.5(i), the coboundary maps correspond to the maps

$$\mathrm{Ker}(\mathrm{N}\colon M'' \to M'')/(s-1)M'' \longrightarrow (M')^G/\mathrm{N}(M'), \quad m'' \mapsto \mathrm{N}(m)\,,$$
$$(M'')^G/\mathrm{N}(M'') \longrightarrow \mathrm{Ker}(\mathrm{N}\colon M' \to M')/(s-1)M', \quad m'' \mapsto (s-1)m\,,$$

where $m \in M$ is an arbitrary element such that $f(m) = m''$.

Hint. Use the resolution from Exercise 1.1.5(i).

EXERCISE 1.1.7. **Cohomology with non-abelian coefficients**

Let G be a group acting by automorphisms of a (possibly non-abelian) group Γ. Let the set $Z^1(G,\Gamma)$ consist of all maps of sets $\varphi\colon G \to \Gamma$ such that the equality

$$\varphi(gh) = \varphi(g) \cdot {}^g\big(\varphi(h)\big)$$

holds for all $g, h \in G$.

(o) Show that for every $\varphi \in Z^1(G,\Gamma)$ one has $\varphi(e) = e$.

(i) Prove that the formula

$$({}^\gamma\varphi)(g) = \gamma \cdot \varphi(g) \cdot {}^g\big(\gamma^{-1}\big), \quad \gamma \in \Gamma, g \in G$$

defines an action of the group Γ on the set $Z^1(G,\Gamma)$. Put

$$H^0(G,\Gamma) = \Gamma^G, \quad H^1(G,\Gamma) = Z^1(G,\Gamma)/\Gamma\,.$$

Note that $H^1(G,\Gamma)$ is a pointed set with a marked element corresponding to the map $\varphi(g) \equiv e$. Show that if G acts trivially on Γ, then there is a canonical bijection between the set $H^1(G,\Gamma)$ and the quotient set of $\mathrm{Hom}(G,\Gamma)$ by Γ (acting by conjugation).

(ii) Let

$$1 \to A \to \widetilde{\Gamma} \to \Gamma \to 1$$

be a short exact sequence of (possibly non-abelian) G-modules. Show that the formula

$$(\delta\gamma)(g) = \tilde{\gamma}^{-1} \cdot {}^{g}\tilde{\gamma}$$

defines a map of sets

$$\delta\colon H^0(G,\Gamma) \to H^1(G,A)\,,$$

where $\gamma \in \Gamma^G \cong H^0(G,\Gamma)$ and $\tilde{\gamma} \in \widetilde{\Gamma}$ is an arbitrary preimage of γ. Prove that there is a short exact sequence

$$1 \to H^0(G,A) \to H^0(G,\widetilde{\Gamma}) \to H^0(G,\Gamma) \xrightarrow{\delta} H^1(G,A) \to H^1(G,\widetilde{\Gamma}) \to H^1(G,\Gamma)$$

of pointed sets, that is, the image of an incoming map coincides with the preimage of the marked element with respect to an outgoing map.

(iii) Assuming the notation of part (ii), suppose that A is a central (and in particular abelian) subgroup in $\widetilde{\Gamma}$. Check that

$$\delta\colon H^0(G,\Gamma) \to H^1(G,A)$$

is a group homomorphism. Show that the formula

$$(\delta\varphi)(g,h) = \tilde{\varphi}(g) \cdot {}^{g}\big(\tilde{\varphi}(h)\big) \cdot \tilde{\varphi}(gh)^{-1}$$

defines a map

$$\delta\colon H^1(G,\Gamma) \to H^2(G,A)$$

of pointed sets, where $\varphi \in Z^1(G,\Gamma)$ and $\tilde{\varphi}(g) \in \widetilde{\Gamma}$ is an arbitrary preimage of the element $\varphi(g)$, $g \in G$. Prove that there is an exact sequence of pointed sets

$$1 \to H^0(G,A) \to H^0(G,\widetilde{\Gamma}) \to H^0(G,\Gamma) \xrightarrow{\delta} H^1(G,A) \to H^1(G,\widetilde{\Gamma}) \to H^1(G,\Gamma) \xrightarrow{\delta} H^2(G,A)\,.$$

(iv) Assuming the notation of part (iii), show that the abelian group $H^1(G,A)$ acts on the set $H^1(G,\widetilde{\Gamma})$ by pointwise multiplication of 1-cocycles, and the kernel of this action coincides with the image $\delta\big(H^0(G,\Gamma)\big)$ of the group $H^0(G,\Gamma)$ with respect to the coboundary map δ. Furthermore, the natural map

$$H^1(G,\widetilde{\Gamma}) \to H^1(G,\Gamma)$$

gives rise to an embedding

$$H^1(G,\widetilde{\Gamma})/H^1(G,A) \hookrightarrow H^1(G,\Gamma)\,.$$

(v) Let G be a cyclic group of order n generated by an element s. Show that the set $Z^1(G,\Gamma)$ is bijective with the set $Z(G,\Gamma)$ of all elements $\alpha \in \Gamma$ such that

$$\alpha \cdot {}^{s}\alpha \cdot {}^{s^2}\alpha \cdot \ldots \cdot {}^{s^{n-1}}\alpha = e\,.$$

Moreover, the action of the group Γ on $Z^1(G,\Gamma)$ defined in part (i) corresponds to the action of Γ on $Z(G,\Gamma)$ given by the formula

$${}^{\gamma}\alpha = \gamma \cdot \alpha \cdot {}^{s}\big(\gamma^{-1}\big)\,.$$

Hint. Map a cocycle $\varphi \in Z^1(G,\Gamma)$ to the element $\alpha = \varphi(s) \in \Gamma$.

Therefore, even for a non-abelian G-module Γ one can compute the first cohomology group as in Exercise 1.1.5.

Exercise 1.1.7(iii) has the following continuation. If the group Γ is abelian, then

$$\delta\colon H^1(G,\Gamma) \to H^2(G,A)$$

is a map between abelian groups. However, in general it may fail to be a group homomorphism. One can show that for all $\varphi, \varphi' \in H^1(G,\Gamma)$ there is an equality in $H^2(G,A)$,

$$\delta(\varphi\varphi') = \delta(\varphi) + \delta(\varphi') + c(\varphi,\varphi'),$$

where the map

$$c\colon H^1(G,\Gamma) \times H^1(G,\Gamma) \to H^2(G,A)$$

is induced by the commutator pairing $\Gamma \times \Gamma \to A$. More details can be found in [**Zar74**].

EXERCISE 1.1.8. **Acyclic resolution**

(i) Given an abelian group A, put $A_+ = \mathrm{Map}(G,A)$. Define a G-module structure on A_+ by the formula

$$({}^g\varphi)(g') = \varphi(g'g), \quad \varphi \in A_+,\ g, g' \in G;$$

cf. Exercise 1.2.4(o). Prove that

$$H^i(G, A_+) = 0$$

for every $i > 0$.

Hint. Compute $H^i(G, A_+)$ via the standard complex from Exercise 1.1.3(iv), using the existence of a natural isomorphism

$$\mathrm{Hom}_G(M, A_+) \cong \mathrm{Hom}(M, A)$$

for every G-module M. The computation also requires Exercise 1.1.3(iii) together with the fact that the abelian groups $\mathbb{Z}[G^{\times i}]$ are free.

(ii) Given a G-module M, consider an embedding of G-modules $M \hookrightarrow M_+$ given by the formula

$$m \mapsto \{g \mapsto {}^g m\}.$$

Put

$$M^0 = M_+, \quad M^1 = (M_+/M)_+, \quad M^2 = \big(M^1/(M_+/M)\big)_+,$$

and proceed like this by induction. Thus we define the G-modules

$$M^i = (C^{i-1})_+, \quad i \geqslant 2,$$

where

$$C^{i-1} = \mathrm{Coker}(M^{i-2} \to M^{i-1})$$

and the map $M^{i-1} \to M^i$ is defined as the composition of the surjection $M^{i-1} \twoheadrightarrow C^{i-1}$ and the natural injection

$$C^{i-1} \hookrightarrow (C^{i-1})_+ = M_i.$$

Check that this gives a complex $M^\bullet$ of G-modules, and show that one has $H^0(M^\bullet) \cong M$, while $H^i(M^\bullet) = 0$ for $i > 0$. This complex is called the *acyclic resolution* of the module M, because $H^i(G, M^j) = 0$ for all $i > 0$ and $j \geqslant 0$.

(iii) Prove that

$$H^i(G, M) \cong H^i\big((M^\bullet)^G\big)$$

for all $i \geqslant 0$.

Hint. Use the acyclicity of the resolution $M^\bullet$ of the module M.

(iv) Show that a short exact sequence

$$0 \to M' \to M \to M'' \to 0$$

of G-modules gives rise to an exact sequence of acyclic resolutions

$$0 \to M'^\bullet \to M^\bullet \to M''^\bullet \to 0.$$

1.2. Behavior under change of group

DEFINITION 1.2.1. Let $f\colon H \to G$ be a group homomorphism. Then f defines a functor

$$f^*\colon \mathrm{Mod}_G \to \mathrm{Mod}_H$$

which maps a G-module M to the same abelian group M with an action of H that comes from f. We usually write M instead of f^*M if it is clear that one has to consider the action of the group H but not of G.

EXERCISE 1.2.2. **Pull-back on cohomology**

(i) Check that f^* is an exact functor.

(ii) Show that the functor f^* defines maps

$$f^*\colon H^i(G, M) \to H^i(H, M), \quad i \geqslant 0\,,$$

for every G-module M.

Hint. Use the exactness of f^*.

(iii) Consider the natural maps

$$f^*\colon \mathrm{Map}(G^{\times i}, M) \to \mathrm{Map}(H^{\times i}, M)\,.$$

Prove that they give a morphism of complexes (see Exercise 1.1.3(v))

$$f^*\colon \mathrm{Map}(G^{\times \bullet}, M) \to \mathrm{Map}(H^{\times \bullet}, M)$$

and that the corresponding maps of the cohomology groups coincide with the maps defined in part (ii).

(iv) Let G be a finite cyclic group generated by an element s, and let $f\colon H \hookrightarrow G$ be an embedding of a subgroup of index r. Fix a generator $t = s^r$ of the subgroup H. Let $\mathrm{N}_G\colon M \to M$ and $\mathrm{N}_H\colon M \to M$ denote the norm maps corresponding to the groups G and H (see Exercise 1.1.5(i)). Show that under the identifications from Exercise 1.1.5(i) with respect to the generators s and t, the homomorphism

$$f^*\colon H^i(G, M) \to H^i(H, M)$$

with i even corresponds to a map

$$M^G/\mathrm{N}_G(M) \longrightarrow M^H/\mathrm{N}_H(M)$$

induced by the identity map from M to itself. Show that for i odd the homomorphism f^* corresponds to a map

$$\mathrm{Ker}(\mathrm{N}_G)/(s-1)M \longrightarrow \mathrm{Ker}(\mathrm{N}_H)/(t-1)M$$

induced by the map

$$M \to M, \quad m \mapsto \sum_{l=0}^{r-1} s^l(m)\,.$$

Hint. Construct a morphism between complexes as in Exercise 1.1.5(i) for the groups G and H that extends the identity map from M to itself in degree zero.

DEFINITION 1.2.3. Let $f\colon H \to G$ be a group homomorphism. Then f defines a functor

$$f_*\colon \mathrm{Mod}_H \to \mathrm{Mod}_G\,, \quad N \mapsto \mathrm{Hom}_H(\mathbb{Z}[G], N)\,,$$

where N is an H-module, the action of H on $\mathbb{Z}[G]$ is given by left multiplication, and the action of G on $\mathrm{Hom}_H(\mathbb{Z}[G], N)$ is given by the formula

$$({}^g\varphi)(g') = \varphi(g'g), \quad \varphi \in \mathrm{Hom}_H(\mathbb{Z}[G], N), \quad g, g' \in G\,.$$

The functor f_* is called *coinduction.*

Note that, in the notation of Definition 1.2.3, there is a canonical isomorphism

$$\mathrm{Hom}_H(\mathbb{Z}[G], N) \cong \mathrm{Map}_H(G, N)$$

of G-modules, where Map_H denotes the set of all H-equivariant maps of sets with an H-action.

EXERCISE 1.2.4. **Coinduction**

(o) Show that $\iota_* A \cong A_+$ for every abelian group A (see Exercise 1.1.8), where

$$\iota\colon \{e\} \hookrightarrow G$$

is the embedding of the trivial subgroup.

(i) Check that Definition 1.2.3 indeed gives a G-action on the abelian group

$$f_*N = \mathrm{Hom}_H(\mathbb{Z}[G], N)\,.$$

(ii) Show that the functor f_* is left-exact.

Hint. Use the fact that the functor $\mathrm{Hom}(\mathbb{Z}[G], -)$ and the functor of H-invariants are left-exact.

(iii) Prove that (f^*, f_*) is a pair of adjoint functors, that is, for every G-module M and every H-module N there is a canonical isomorphism

$$\mathrm{Hom}_H(M, N) \cong \mathrm{Hom}_G(M, f_*N)\,.$$

Hint. Adapt a proof of the similar assertion for modules over commutative rings. In particular, send a map $\psi \in \mathrm{Hom}_H(M, N)$ to the map

$$m \mapsto \{g \mapsto {}^g(\psi(m))\}$$

and, conversely, send a map $\xi \in \mathrm{Hom}_G(M, f_*N)$ to the map

$$m \mapsto \xi(m)(e)\,,$$

where $e \in \mathbb{Z}[G]$ is the neutral element of the group and of the group ring.

(iv) Show that for a composition $f \circ g$ of group homomorphisms there is a canonical isomorphism of functors

$$f_* \circ g_* \cong (f \circ g)_* .$$

EXERCISE 1.2.5. **Coinduction for an injective homomorphism**
Let $f\colon H \hookrightarrow G$ be an injective group homomorphism, and let N be an H-module.

(i) Note that there is a (non-canonical) isomorphism

$$f_*N \cong \prod_{H\backslash G} N$$

of abelian groups, and use it to deduce that the functor f_* is exact.

(ii) Prove that every projective G-module M is also projective as an H-module.

Hint. Use part (i) and Exercise 1.2.4(iii).

(iii) Prove *Shapiro's lemma*: for every H-module N there is a canonical isomorphism

$$H^i(H, N) \cong H^i(G, f_*N), \quad i \geqslant 0 .$$

Hint. Use part (ii) and the fact that cohomology groups can be computed via any resolution of the module $\mathbb{Z}$. Thus the required isomorphism comes from Exercise 1.2.4(iii). Another way is to use Exercise 1.1.8(iii), the isomorphism

$$f_*(N^\bullet) \cong (f_*N)^\bullet$$

of acyclic resolutions provided by Exercise 1.2.4(o) and (iv), and the isomorphism

$$(f_*N)^G \cong N^H$$

from Exercise 1.2.4(iii).

(iv) Suppose that the subgroup H is normal in G, and choose representatives $\{g_i\}$ of right cosets $H \setminus G$. We get automorphisms

$$g_i\colon H \to H, \quad h \mapsto g_i h g_i^{-1} .$$

Take an element $\varphi \in \mathrm{Hom}_H(\mathbb{Z}[G], N)$, and assign to it a collection

$$\{\varphi(g_i)\} \in \prod_i g_i^* N .$$

Check that there is an isomorphism

$$f^* f_* N \cong \prod_i g_i^* N$$

of H-modules.

(v) Show that

$$H^i(H, N_+) = 0$$

for every $i > 0$, where

$$N_+ = \mathrm{Map}(G, N) .$$

Hint. Use part (iv), Exercise 1.2.4(o) and (iv), and Exercise 1.1.8(i).

If the reader is unfamiliar with spectral sequences, we advise skipping the next exercise except for its part (i). We will not make much use of these results in what follows.

EXERCISE 1.2.6. **Coinduction for a surjective homomorphism**
Let $f\colon H \twoheadrightarrow G$ be a surjective group homomorphism with kernel I, and let N be an H-module.

(i) Show that there is an isomorphism of G-modules
$$f_*N \cong N^I .$$

(ii) Check that
$$H^i\big(G, (N_+)^I\big) = 0$$
for every $i > 0$, where
$$N_+ = \mathrm{Map}(H, N) .$$
Hint. Use part (i), and then proceed similarly to Exercise 1.2.5(v), namely, apply Exercises 1.2.4(o),(iv) and 1.1.8(i).

(iii) Show that a group $H^j(I, N)$ has a canonical structure of a G-module, and prove that there is a spectral sequence with the second page
$$E_2^{ij} = H^i\big(G, H^j(I, N)\big) \tag{1.2}$$
that converges to $H^{i+j}(H, N)$.

Hint. Start with the acyclic resolution $N^\bullet$ of N with respect to the group H (see Exercise 1.1.8 with $G = H$ and $M = N$), and consider the complex $(N^\bullet)^I$ of G-modules. According to Exercise 1.2.5(v), its cohomology groups are isomorphic to $H^j(I, N)$, which gives an action of G on $H^j(I, N)$. Now consider the bi-complex
$$C^{\bullet,\bullet} = \big((N^\bullet)^I\big)^\bullet$$
obtained by taking the acyclic resolution of terms of the complex $(N^\bullet)^I$ with respect to the group G. Finally, use two spectral sequences related to the bi-complex $\big(C^{\bullet,\bullet}\big)^G$. By part (ii), one of these spectral sequences degenerates to a single line equal to $(N^\bullet)^G$, and the other has the form (1.2) by Exercises 1.2.5(v) and 1.1.8(iv). Note that this is a particular case of a spectral sequence associated with a composition of derived functors, sometimes called a Leray spectral sequence.

(iv) Show that there is an exact sequence
$$0 \to H^1(G, N^I) \to H^1(H, N) \to H^1(I, N)^G \to H^2(G, N^I) \to H^2(H, N) .$$
Hint. Use the spectral sequence from part (iii).

EXERCISE 1.2.7. **Direct image on cohomology**
Let $f\colon H \hookrightarrow G$ be an embedding of a subgroup such that H has a finite index in G.

(i) Prove that (f_*, f^*) is a pair of adjoint functors, that is, for every G-module M and every H-module N there is a canonical isomorphism

$$\mathrm{Hom}_G(f_*N, M) \cong \mathrm{Hom}_H(N, M)\,.$$

Hint. Send a map $\psi \in \mathrm{Hom}_H(N, M)$ to the map

$$\varphi \mapsto \sum_{[g]\in G/H} {}^g\big(\psi(\varphi(g^{-1}))\big),$$

where the sum is taken over representatives of cosets. In the opposite direction, send a map $\xi \in \mathrm{Hom}_G(f_*N, M)$ to the map

$$n \mapsto \xi(\chi_n)\,,$$

where $\chi_n \in \mathrm{Hom}_H(\mathbb{Z}[G], N)$ is defined by the formula

$$\chi_n(h) = {}^h n$$

for $h \in H$ and $\chi_n(g) = 0$ for $g \notin H$. Use the fact that for every homomorphism $\varphi \in \mathrm{Hom}_H(\mathbb{Z}[G], N)$ one has

$$\varphi = \sum_{[g]\in G/H} {}^g\big(\chi_{\varphi(g^{-1})}\big)\,,$$

where the sum is taken over representatives of cosets.

Therefore, under the assumption of the exercise, the functor f^* is both a left adjoint and a right adjoint to f_*; cf. Exercise 1.2.4(iii).

(ii) Using part (i), for any G-module M construct a canonical morphism of G-modules

$$f_*f^*M \to M\,,$$

and define the direct image on cohomology groups

$$f_*\colon H^i(H, M) \to H^i(G, M)$$

by applying Shapiro's lemma (Exercise 1.2.5(iii)).

(iii) Prove that the composition

$$f_* \circ f^*\colon H^i(G, M) \to H^i(G, M)$$

is the same as multiplication by the index of H in G.

Hint. Start with an explicit computation for $i = 0$, and then deduce the general case by induction on i using a long exact sequence of cohomology groups associated with the short exact sequence

$$0 \to M \to M_+ \to M_+/M \to 0\,;$$

see Exercise 1.1.8. Do not forget to check that f_* and f^* are compatible with coboundary maps.

Note that sometimes the following terminology is used. Given an embedding $f\colon H \hookrightarrow G$ of a subgroup, the pull-back map

$$f^*\colon H^i(G, M) \to H^i(H, M)$$

is denoted by Res and is called the *restriction.* If the index of H in G is finite, then the direct image map

$$f_*\colon H^i(H, M) \to H^i(G, M)$$

is denoted by Cor and is called the *corestriction*. In this case one has

$$\mathrm{Cor} \circ \mathrm{Res} = [G : H]$$

by Exercise 1.2.7(iii). If a homomorphism $f\colon H \to G$ is surjective and has kernel I, then the composition of the pull-back map

$$f^*\colon H^i(G, M^I) \to H^i(H, M^I)$$

with the natural map $H^i(H, M^I) \to H^i(H, M)$ is called the *inflation* and is denoted by Inf. There is an exact sequence

$$0 \to H^1(G, N^I) \xrightarrow{\mathrm{Inf}} H^1(H, N) \xrightarrow{\mathrm{Res}} H^1(I, N)\,;$$

see Exercise 1.2.6(iv).

1.3. Cohomology of finite groups

EXERCISE 1.3.1. **Finiteness of cohomology groups**
Let G be a finite group.

(i) Prove that for every G-module M, the cohomology groups $H^i(G, M)$ are annihilated by multiplication by the order $|G|$ when $i > 0$.

Hint. Apply Exercise 1.2.7(iii) to the map $\{e\} \hookrightarrow G$.

(ii) Prove that for every finitely generated G-module M, the cohomology groups $H^i(G, M)$ are finite when $i > 0$.

Hint. Use the standard complex to show that the groups $H^i(G, M)$ are finitely generated, and then apply part (i).

EXERCISE 1.3.2. **Restriction to a Sylow subgroup**
Let G be a finite group, M a finitely generated G-module, and i a positive integer. Given a prime number p, let G_p be a Sylow p-subgroup in G (in particular, if p does not divide the order of the group G, then G_p is trivial).

(i) Let p be a prime number. Let H_p be the largest p-primary subgroup of the group $H^i(G, M)$; it is well-defined because the group $H^i(G, M)$ is a finite abelian group by Exercise 1.3.1(ii). Show that the restriction map

$$H_p \to H^i(G_p, M)$$

is injective.

Hint. Consider the composition

$$H_p \to H^i(G_p, M) \to H^i(G, M)$$

and apply Exercise 1.2.7(iii) to the embedding $G_p \hookrightarrow G$.

(ii) Suppose that for every prime p and every Sylow subgroup $G_p \subset G$ one has $H^i(G_p, M) = 0$. Prove that $H^i(G, M) = 0$.

Hint. Use part (i).

(iii) Does the converse assertion to part (ii) hold?

Hint. Consider the group $G = \mathrm{S}_3$ and the trivial G-module $M = \mathbb{Z}/3\mathbb{Z}$.

EXERCISE 1.3.3. **Cohomology with coefficients in $\mathbb{Q}/\mathbb{Z}$**
Let G be a finite group.

(i) Let M be an arbitrary G-module which is also a vector space over $\mathbb{Q}$. Prove that
$$H^i(G, M) = 0$$
for every $i > 0$.

Hint. Use Exercise 1.3.1(i).

(ii) Prove that $H^1(G, \mathbb{Z}) = 0$ and there is a canonical isomorphism
$$H^2(G, \mathbb{Z}) \cong \mathrm{Hom}(G, \mathbb{Q}/\mathbb{Z}),$$
where $\mathbb{Z}$ and $\mathbb{Q}/\mathbb{Z}$ are treated as trivial G-modules.

Hint. Use the exact sequence of G-modules
$$0 \to \mathbb{Z} \to \mathbb{Q} \to \mathbb{Q}/\mathbb{Z} \to 0$$
together with part (i).

(iii) Let G be a cyclic group of order n generated by an element s. Let
$$\theta\colon \mathbb{Z}/n\mathbb{Z} \xrightarrow{\sim} \mathrm{Hom}(G, \mathbb{Q}/\mathbb{Z})$$
be a composition of isomorphisms from part (ii) and Exercise 1.1.5(i). Show that θ maps the element $1 \in \mathbb{Z}/n\mathbb{Z}$ to the homomorphism
$$G \to \mathbb{Q}/\mathbb{Z}, \quad s \mapsto [1/n].$$

Hint. Use Exercises 1.1.5(ii) and 1.1.6(iii).

1.4. Permutation and stably permutation modules

DEFINITION 1.4.1. Let G be an arbitrary group. Let M be a G-module which is free as a $\mathbb{Z}$-module. We say that M is a *permutation* G-module if M has a basis over $\mathbb{Z}$ such that the group G acts by permutations of its elements.

EXERCISE 1.4.2. **Cohomology with coefficients in a permutation module**

Let G be a finite group, M a permutation G-module, B a basis in M such that G permutes its elements, and $\mathbb{O}$ the set of G-orbits in B.

(i) Show that M can be written as
$$M \cong \bigoplus_{O \in \mathbb{O}} \mathbb{Z}\big[G/\operatorname{Stab}_G(x_O)\big],$$
where x_O is (any) element from an orbit O, the subgroup $\operatorname{Stab}_G(x_O)$ is the stabilizer of x_O in the group G, and $G/\operatorname{Stab}_G(x_O)$ is the set of cosets of this subgroup.

(ii) Prove that $H^1(G, M) = 0$ and
$$H^2(G, M) \cong \bigoplus_{O \in \mathbb{O}} \mathrm{Hom}\left(\operatorname{Stab}_G(x_O), \mathbb{Q}/\mathbb{Z}\right).$$

Hint. Use part (i) and the fact that for any embedding $i\colon H \hookrightarrow G$ of a subgroup there is an isomorphism of G-modules
$$i_*\mathbb{Z} \cong \mathbb{Z}[G/H].$$

Then apply Shapiro's lemma (see Exercise 1.2.5(iii)) together with Exercise 1.3.3(ii).

DEFINITION 1.4.3. Let G be an arbitrary group, and let M be a G-module. We say that M is a *stably permutation* G-module if for some permutation G-modules N_1 and N_2 there is an isomorphism of G-modules

$$M \oplus N_1 \cong N_2 \, .$$

EXERCISE 1.4.4. **Restriction of a stably permutation module**

Let G be an arbitrary group, and let M be a (stably) permutation G-module. Let $f\colon H \to G$ be a group homomorphism. Show that f^*M is a (stably) permutation H-module. In particular, a restriction of a (stably) permutation module to a subgroup is again a (stably) permutation module.

EXERCISE 1.4.5. **Cohomology with coefficients in a stably permutation module**

Let G be a finite group, and let M be a stably permutation G-module.

(i) Prove that

$$H^1(G, M) = 0 \, .$$

Hint. Use Exercise 1.4.2(ii).

(ii) Using part (i) and Exercise 1.4.4, show that the module M satisfies the following condition:

$$\text{for every subgroup } H \subset G, \text{ one has } H^1(H, M) = 0. \tag{1.3}$$

(iii) Let N be an arbitrary G-module such that its restriction to every Sylow subgroup in G is a stably permutation module. Prove that N satisfies condition (1.3).

Hint. Use part (ii) and Exercise 1.3.2(ii).

CHAPTER 2

Galois Cohomology

Let $K \subset L$ be a finite Galois extension of fields. Cohomology of the group $\mathrm{Gal}(L/K)$ is called *Galois cohomology*. We will use the notation

$$G = \mathrm{Gal}(L/K)$$

in §2.1 and §2.2.

2.1. Descent for fibered categories

Galois cohomology is closely related to the following informal questions. Suppose that we have some algebro-geometric object Y defined over a field L (see Example 2.1.1 below). What conditions are sufficient to imply that Y is actually defined over a smaller field K, that is, there is some similar object X over K, and Y is obtained from X by scalar extension from K to L? How can one describe all objects X that solve this problem? The following definitions, examples, and exercises are devoted to formalization of these questions and their interpretation in terms of Galois cohomology.

Suppose that for every field E one has fixed a category $\mathcal{M}(E)$ and for every field extension $E \subset F$ one has fixed a functor

$$\mathcal{M}(E) \to \mathcal{M}(F), \quad X \mapsto X_F\,.$$

This functor is usually referred to as *scalar extension*. Moreover, suppose that for every tower of fields $E \subset F \subset M$ and every object X in $\mathcal{M}(E)$ one has fixed an isomorphism

$$(X_F)_M \cong X_M$$

that is functorial with respect to X and behaves well in towers of the form

$$E \subset F \subset M \subset N\,.$$

Such a collection $\mathcal{M}$ of categories $\mathcal{M}(E)$ is called a *category fibered over fields*. In this chapter we will also call it a *fibered category* for simplicity. In what follows $\mathcal{M}$ will denote a fibered category.

EXAMPLE 2.1.1. *Categories fibered over fields*

The following examples of categories $\mathcal{M}(E)$ provide categories fibered over fields; the functors $\mathcal{M}(E) \to \mathcal{M}(F)$ are defined as tensor products with F over E:

(i) the category of vector spaces (including infinite-dimensional ones) over a field E;
(ii) the category of vector spaces with quadratic form over E;
(iii) the category of associative algebras over E;
(iv) the category of affine varieties over E;
(v) the category of quasi-projective varieties over E;

(vi) the category of quasi-projective varieties over E with marked E-point;
(vii) the category of algebraic groups over E.

Recall that $K \subset L$ denotes a finite Galois extension with Galois group G. Every element $g \in G$ defines an isomorphism $g\colon L \xrightarrow{\sim} L$, which can also be considered as an extension of fields. Since $\mathcal{M}$ is a fibered category, this gives a functor

$$g_*\colon \mathcal{M}(L) \to \mathcal{M}(L)\,.$$

Thus, for every object Y in $\mathcal{M}(L)$ there is a well-defined object g_*Y in $\mathcal{M}(L)$, and for every morphism $\theta\colon Y \to Y'$ there is a well-defined morphism

$$g_*\theta\colon g_*Y \to g_*Y'\,.$$

One can consider such a collection of data as an action of the group G on the category $\mathcal{M}(L)$.

If $Y \cong X_L$ for some object X in $\mathcal{M}(K)$, then the assumption on extension of scalars in the tower of fields

$$K \hookrightarrow L \overset{g}{\hookrightarrow} L$$

gives a canonical isomorphism

$$g_*X_L \xrightarrow{\sim} X_L\,.$$

One can see that the isomorphism

$$(gh)_*X_L \xrightarrow{\sim} X_L$$

coincides with the composition of isomorphisms $g_*(h_*X_L \xrightarrow{\sim} X_L)$ and $g_*X_L \xrightarrow{\sim} X_L$ for all $g, h \in G$. For all objects X and X' in $\mathcal{M}(K)$, the group G acts on the set of morphisms from X_L to X'_L in the category $\mathcal{M}(L)$. Specifically, for every morphism

$$\theta\colon X_L \to X'_L$$

the result of the action of g on θ, which we denote, as usual, by ${}^g\theta$, is defined by the condition that the following diagram is commutative:

$$\begin{array}{ccc} g_*X_L & \xrightarrow{\sim} & X_L \\ \big\downarrow{\scriptstyle g_*\theta} & & \big\downarrow{\scriptstyle {}^g\theta} \\ g_*X'_L & \xrightarrow{\sim} & X'_L \end{array} \tag{2.1}$$

In particular, the automorphism group $\mathrm{Aut}(X_L)$ of an object X_L of the category $\mathcal{M}(L)$ is always a (possibly non-abelian) G-module.

We will say that one has *descent data* on an object Y of a category $\mathcal{M}(L)$ if for every element $g \in G$ one has fixed an isomorphism $\rho(g)\colon g_*Y \xrightarrow{\sim} Y$ satisfying the above condition on composition of elements from G; that is, we require that

$$\rho(gh) = \rho(g) \circ g_*(\rho(h))$$

for all $g, h \in G$. One can think of Y as being a G-equivariant object in $\mathcal{M}(L)$.

In particular, as we saw above, for every object X in $\mathcal{M}(K)$ one always has canonical descent data on the object X_L. We will denote them by

$$\rho_X(g)\colon g_*X \xrightarrow{\sim} X\,.$$

Morphisms between objects with descent data (Y, ρ) and (Y', ρ') are defined in a natural way: these are morphisms $\theta\colon Y \to Y'$ such that for every $g \in G$ one has an equality

$$\theta \circ \rho(g) = \rho'(g) \circ g_*\theta$$

between morphisms from g_*Y to Y'. Informally speaking, θ commutes with g. We will denote the category of objects in $\mathcal{M}(L)$ with descent data by $\mathcal{M}(L)^G$.

An isomorphism between objects with descent data is a morphism of objects with descent data that has an inverse. In particular, descent data ρ and ρ' on the same object Y are considered to be isomorphic if there is an automorphism σ of the object Y providing an isomorphism of objects with descent data,

$$\sigma\colon (Y, \rho) \to (Y, \rho').$$

We will say that $\mathcal{M}$ *satisfies descent* if for every finite Galois extension $K \subset L$ the functor

$$X \mapsto (X_L, \rho_X)$$

from $\mathcal{M}(K)$ to $\mathcal{M}(L)^G$ is an equivalence of categories. In other words, two conditions have to be satisfied. First, for every two objects X and X' in $\mathcal{M}(K)$ the natural map between sets of morphisms of objects in the corresponding categories,

$$\mathrm{Hom}(X, X') \to \mathrm{Hom}\big((X_L, \rho_X), (X'_L, \rho_{X'})\big),$$

is bijective. Second, for every object (Y, ρ) in $\mathcal{M}(L)^G$ there exists an object X in $\mathcal{M}(K)$ such that there is an isomorphism of objects with descent data

$$(Y, \rho) \cong (X_L, \rho_X).$$

In particular, the object X is uniquely defined up to an isomorphism in $\mathcal{M}(K)$. Such an object X will be called the *descent* of (Y, ρ) from L to K. For simplicity we will sometimes omit ρ and call X just the descent of Y. Also, we will usually need to check only the second of the two conditions above, since the first one comes for free in most cases.

Given an object X in $\mathcal{M}(K)$, by an *L-form* of X we mean an object X' in $\mathcal{M}(K)$ such that there is an isomorphism $X_L \cong X'_L$ in the category $\mathcal{M}(L)$. Two L-forms of X are said to be isomorphic if they are isomorphic in the category $\mathcal{M}(K)$. A *form* of an object X is its L-form for some finite Galois extension $K \subset L$. We denote the set of isomorphism classes of L-forms of an object X by $\Phi(X, L)$, and we denote the set of isomorphism classes of forms of X over all possible fields by $\Phi(X)$.

EXAMPLE 2.1.2. *Descent data on fibered categories*

(i) Let $\mathcal{M}$ be the fibered category of vector spaces, and let U be a vector space over L. Then g_*U coincides with U as an abelian group, but the action of the field L on g_*U is defined by the formula

$$\lambda \cdot u = \big(g^{-1}(\lambda)\big)u$$

for all $\lambda \in L$ and $u \in U$. If V is a vector space over K, then the canonical isomorphism $g_*V_L \xrightarrow{\sim} V_L$ (recall that $V_L = L \otimes_K V$) is given by the formula

$$\lambda \otimes v \mapsto g(\lambda) \otimes v.$$

After a choice of basis of V over K, the isomorphism $g_*V_L \xrightarrow{\sim} V_L$ is given by applying g to coordinates of vectors. The action of G on the automorphism group $\mathrm{Aut}(V_L) \cong \mathrm{GL}(V_L)$ is given by applying g to matrix elements of L-linear operators. Fixing descent data on the vector space U over L amounts to fixing a *semilinear action* of G on U, that is, an action with the property

$$g(\lambda u) = g(\lambda)g(u)$$

for all $g \in G$, $\lambda \in L$, and $u \in U$. For instance, such a situation arises when one has chosen a K-linear representation of the group G on a vector space V over K and takes $U = V_L$ (cf. Exercise 2.2.1(i) below).

(ii) Let $\mathcal{M}$ be the fibered category of vector spaces with quadratic form, let U be a vector space over L, and let $q\colon U \to L$ be a quadratic form. Then the pair $g_*(U, q)$ consists of the vector space g_*U and the quadratic form defined by the formula

$$(g_*q)(u) = g\big(q(u)\big)$$

for $u \in U$. Given a vector space with a quadratic form (V, p) over K, the isomorphisms

$$g_*(V_L, p_L) \xrightarrow{\sim} (V_L, p_L)$$

and the action of the group G on $\mathrm{Aut}(V_L, p_L)$ are defined in the same way as in part (i). Note that the action of the group G on matrix elements of L-linear operators from V_L to itself preserves the orthogonal subgroup $\mathrm{O}(V_L, p_L)$ in $\mathrm{GL}(V_L)$. Fixing descent data on an object (U, q) from $\mathcal{M}(L)$ is equivalent to fixing a semilinear action of G on U such that

$$g\big(q(u)\big) = q\big(g(u)\big)$$

for all $g \in G$ and $u \in U$.

(iii) Let $\mathcal{M}$ be the fibered category of associative algebras, and let B be an associative algebra over L. Then g_*B has the structure of a vector space described in part (i). In particular, g_*B coincides with B as an abelian group. The multiplication map

$$g_*B \times g_*B \to g_*B$$

is the same as in B. For an algebra A over K, the isomorphisms

$$g_*A_L \xrightarrow{\sim} A_L$$

and the action of G on A_L are defined in the same way as in part (i). Fixing descent data on an algebra B over L is equivalent to fixing a semilinear action of G on B that commutes with multiplication in B.

(iv) Let $\mathcal{M}$ be the fibered category of affine varieties, let Y be an affine variety over L, and let B be the L-algebra of regular functions on Y, so that $Y \cong \mathrm{Spec}(B)$. Then the algebra of regular functions on the variety g_*Y is defined as g_*B, and everything goes as in part (iii). Note that g_*Y is canonically isomorphic over L to the variety whose equations are obtained by applying g to equations of Y.

(v) Let $\mathcal{M}$ be the fibered category of quasi-projective varieties, and let Y be a quasi-projective variety over L. Write $Y = Y_1 \setminus Y_2$, where Y_1 and Y_2 are projective varieties in a projective space $\mathbb{P}^n$. Then $g_*Y = g_*Y_1 \setminus g_*Y_2$, and each g_*Y_i is a projective variety in $\mathbb{P}^n$ whose equations are obtained by applying g to equations of Y_i. For a quasi-projective variety X over K, the isomorphism

$$g_*X_L \xrightarrow{\sim} X_L$$

corresponds to the identity map from X to itself. The action of G on an automorphism in $\mathrm{Aut}(X_L)$ is given by applying g to coefficients of homogeneous polynomials on $\mathbb{P}^n$ that define (locally on X_L) the given automorphism.

(vi) Let $\mathcal{M}$ be the fibered category of quasi-projective varieties with marked point, and let (Y, y) be a quasi-projective variety over L with a point $y \in Y$ that has coordinates from L. Then $g_*(Y, y)$ is the variety g_*Y with the point g_*y whose coordinates are obtained by applying g to coordinates of the point y. All the rest is defined as in part (v).

(vii) Let $\mathcal{M}$ be the fibered category of algebraic groups, and let H be an algebraic group over L. Then g_*H has the structure of a variety over L described in part (v), the unit in g_*H is defined as in part (vi), and the multiplication and inversion morphisms are obtained by applying g to coefficients of the corresponding morphisms for the group H. All the rest is defined as in part (v).

Many important algebro-geometric objects form fibered categories that satisfy descent. In particular, we will see in Exercises 2.1.3 and 2.1.4 that this is the case for all fibered categories from Example 2.1.1. Moreover, in all these cases uniqueness of the object X that solves the descent problem for descent data (Y, ρ) is obvious, so that it only remains to check that X exists.

EXERCISE 2.1.3. **Descent for vector spaces**

(i) Let (U, ρ) be a vector space over L with descent data, that is, with a semilinear action ρ of the group G on U; see Example 2.1.2(i). Put $V = U^G$. Construct a canonical map of vector spaces over L with descent data

$$(V_L, \rho_V) \to (U, \rho),$$

that is, an L-linear map $\alpha\colon V_L \to U$ commuting with the (semilinear) action of G.

(ii) Let $K \subset E$ be an arbitrary field extension. Consider the L-algebra

$$A = E \otimes_K L$$

with an action of G via the right factor L. Also consider the A-module

$$W = A \otimes_L U \cong E \otimes_K U$$

with an A-semilinear action of the group G via the right factor U. Check that

$$A^G = E \otimes 1 \cong E.$$

Show that the L-linear map

$$\alpha\colon L \otimes_K U^G \to U$$

defined in part (i) is an isomorphism if and only if the natural homomorphism of A-modules

$$\beta\colon A \otimes_E W^G \to W$$

is an isomorphism.

Hint. Use the isomorphism $W^G \cong E \otimes_K U^G$.

(iii) Suppose that $E = L$ or $E = K^{sep}$. Prove that there is a G-equivariant isomorphism of rings

$$A \cong \prod_G E\,,$$

where the group G acts on $\prod_G E$ by permutation of factors.

Hint. Put $L = K(\theta)$, and let $f \in K[T]$ be the minimal polynomial of the element $\theta \in L$. Then f splits into a product of linear factors over the field E.

(iv) Under the assumptions of part (iii), prove that every G-equivariant A-module has the form $\bigoplus_G P$, where P is an E-vector space, the group G acts by permutation of summands, and the structure of an A-module comes from the isomorphism $A \cong \prod_G E$ from part (iii).

(v) Under the assumptions of part (iii), prove that for every G-equivariant A-module M, the natural homomorphism of A-modules

$$A \otimes_E M^G \to M$$

is an isomorphism.

Hint. Apply part (iv).

Use this together with part (ii) to deduce that the map α is an isomorphism. Therefore, the fibered category of vector spaces satisfies descent.

More explicitly, Exercise 2.1.3 means that every L-vector space with a semilinear action of the group G has a G-invariant basis.

EXERCISE 2.1.4. **Other examples of descent**

(i) Prove that the fibered category of vector spaces with a quadratic form satisfies descent.

Hint. Let (U, q) be a vector space with quadratic form over L. Suppose that one has fixed descent data on (U, q); see Example 2.1.2(ii). Construct a vector space V over K as in Exercise 2.1.3, and define a quadratic form p on V as the restriction of q to $V \subset V_L \cong U$.

(ii) Prove that the fibered category of associative algebras satisfies descent.

Hint. Do the same as in part (i).

(iii) Prove that the fibered category of affine varieties satisfies descent.

Hint. This is a particular case of part (ii).

(iv) Prove that the fibered category of quasi-projective varieties satisfies descent.

Hint. A quasi-projective variety Y over L can be considered as a quasi-projective variety over K obtained as the composition of morphisms

$$Y \to \mathrm{Spec}(L) \to \mathrm{Spec}(K)\,.$$

Given descent data ρ on Y (see Example 2.1.2(v)), define the action of an element $g \in G$ on the variety Y over K as the composition $\rho(g) \circ g$, where

$$g\colon Y \to g_*Y$$

acts by applying g to coordinates of points in Y. Since Y is quasi-projective, there is an atlas on Y considered as a variety over K that consists of open affine G-invariant subsets. This allows us to reduce everything to part (iii). Therefore, the descent of the variety Y is the quotient Y/G by the above action of the group G, where Y is considered as a variety over the field K.

(v) Prove that the fibered category of quasi-projective varieties with marked point satisfies descent.

Hint. Let (Y, y) be a quasi-projective variety with a marked point defined over L. Suppose that one has fixed descent data on (Y, y); see Example 2.1.2(vi). Construct a quasi-projective variety X over K as in part (iv). Note that the point y corresponds to a point on the variety $X_L \cong Y$ whose coordinates are invariant under the action of G; this defines a K-point x on X.

(vi) Prove that the fibered category of algebraic groups satisfies descent.

Hint. Since algebraic groups are quasi-projective varieties, one can do the same as in part (v).

EXERCISE 2.1.5. **Descent for projective varieties**

Let Y be a variety over L with descent data ρ, and let X be the descent of Y from L to K. Let U be a finite-dimensional L-vector space with descent data, that is, one has fixed a semilinear action of the group G on U; see Example 2.1.2(i).

(o) Show that the descent of the projective space $\mathbb{P}(U)$ over L is the projective space $\mathbb{P}(U^G)$ over K.

Hint. Using Exercise 2.1.3(ii) and (v), construct an isomorphism

$$\mathbb{P}(U^G)_L \cong \mathbb{P}(U)$$

of varieties with descent data over L.

(i) Write $\dim(U) = m + 1$. Suppose that the closed embedding

$$\iota \colon Y \hookrightarrow \mathbb{P}^m = \mathbb{P}(U)$$

over L is a morphism of varieties with descent data. Show that there is a closed embedding

$$X \hookrightarrow \mathbb{P}^m = \mathbb{P}(U^G)$$

over K.

Hint. Use part (o).

(ii) Under the assumptions of part (i), let $K \subset F$ be an arbitrary field extension. Choose a non-zero vector $v \in (U^G)_F$ and consider the corresponding F-point

$$[v] \in \mathbb{P}(U^G)(F)\,.$$

Let

$$\phi \colon (U^G)_F \to U_F$$

be the extension of scalars from K to F of the natural embedding of K-vector spaces $U^G \to U$. Show that $[v] \in X(F)$ if and only if

$$[\phi(v)] \in Y(F)\,.$$

Hint. This is immediately implied by the isomorphism $X_K \cong Y$.

(iii) Let $Y \subset \mathbb{P}^n = \mathbb{P}(W)$ be a closed embedding over L, where W is an L-vector space of dimension $n+1$. Prove that there is a closed embedding

$$X \subset \mathbb{P}^{(n+1)^d-1}$$

over K, where $d = [L : K]$. Thus, the descent of a projective variety is a projective variety as well.

Hint. Consider the vector space

$$U = \bigotimes_{g\in G} g_*W\,;$$

see Example 2.1.2(i). Define descent data on U as the (semilinear) action of the group G given on decomposable tensors by the formula

$$h\colon \bigotimes_g w_g \mapsto \bigotimes_g w_{h^{-1}g}\,.$$

Define an embedding $\iota\colon Y \hookrightarrow \mathbb{P}(U)$ as the composition of the embedding

$$\prod_g \rho(g)^{-1}\colon Y \hookrightarrow \prod_{g\in G} g_*Y\,,$$

the natural map

$$\prod_{g\in G} g_*Y \hookrightarrow \prod_{g\in G} \mathbb{P}(g_*W)\,,$$

and the Segre embedding

$$\prod_{g\in G} \mathbb{P}(g_*W) \hookrightarrow \mathbb{P}(U)\,.$$

Show that ι is a morphism of objects with descent data. Finally, apply part (i).

2.2. Forms and first Galois cohomology

In this section $\mathcal{M}$ denotes a category fibered over fields.

EXERCISE 2.2.1. **Descent data, 1-cocycles, and forms**

(i) Show that for every object X in $\mathcal{M}(K)$ there is a canonical bijection between $H^1\big(G, \mathrm{Aut}(X_L)\big)$ (see Exercise 1.1.7) and the set of isomorphism classes of descent data on X_L; this bijection sends the marked element in $H^1\big(G, \mathrm{Aut}(X_L)\big)$ to the canonical descent data ρ_X on X_L.

Hint. Use the correspondence

$$\rho(g) = \varphi(g) \circ \rho_X(g)$$

between descent data ρ and 1-cocycles φ. Also recall that

$${}^g\theta \circ \rho_X(g) = \rho_X(g) \circ g_*\theta$$

for every isomorphism $\theta\colon X_L \to X_L$; see diagram (2.1). In particular, every homomorphism $G \to \mathrm{Aut}(X)$ gives descent data on X_L; see Exercise 1.1.4(i).

(ii) Let X be an object in $\mathcal{M}(K)$ and X' its L-form. Then X' defines descent data $\rho_{X'}$ on the object $X_L \cong X'_L$ that are possibly different from the descent data ρ_X. Prove that this gives a map from the set $\Phi(X, L)$ of isomorphism classes of L-forms of the object X to the set of isomorphism classes of descent data on X_L.

(iii) Show that if $\mathcal{M}$ satisfies descent, then there is a canonical bijection between the set $\Phi(X, L)$ and the set of isomorphism classes of descent data on X_L. Therefore, in this case part (i) gives a canonical bijection

$$\Phi(X, L) \cong H^1\big(G, \mathrm{Aut}(X_L)\big)\,.$$

In particular, a 1-cocycle

$$\varphi \in Z^1\big(G, \mathrm{Aut}(X_L)\big)$$

defines a form of the variety X called the *twist of X by φ*.

(iv) Let X be a variety that consists of two points defined over K. Describe all forms of X.

(v) Let $R \subset L$ be the set of roots of an irreducible polynomial that defines the extension $K \subset L$. Show that $\mathrm{Spec}(L)$ is an L-form of the variety

$$X = \coprod_R \mathrm{Spec}(K)$$

and that the corresponding 1-cocycle is given by the natural homomorphism

$$G \to \mathrm{Aut}(X) = \mathrm{Aut}(R)\,;$$

cf. part (i).

(vi) Prove *Hilbert's Theorem 90*:

$$H^1\big(G, \mathrm{GL}_n(L)\big) = \{1\}\,.$$

Hint. By Exercise 2.1.3 the category of (finite-dimensional) vector spaces satisfies descent. Since all vector spaces of the same dimension are isomorphic to each other, it remains to apply part (iii).

In particular, for $n = 1$ we obtain

$$H^1(G, L^*) = \{1\}\,.$$

There is also an explicit algebraic proof of Hilbert's Theorem 90.

EXERCISE 2.2.2. **Hilbert's Theorem 90 (algebraic proof)**

(i) Prove *Artin's lemma on independence of characters*: let

$$\chi_1, \ldots, \chi_n \colon H \to L^*$$

be pairwise different homomorphisms from an arbitrary group H to L^*. Then $\chi_1, \ldots, \chi_n$ are independent over L; that is, if for some $\lambda_1, \ldots, \lambda_n \in L$ one has

$$\sum_{i=1}^{n} \lambda_i \chi_i(h) = 0$$

for every $h \in H$, then all λ_i equal zero.

Hint. Consider a minimal linear dependence

$$\sum_{i=1}^{n} \lambda_i \chi_i = 0.$$

It yields $\sum_{i=1}^{n} \lambda_i \chi_i(gh) = 0$ for arbitrary $g, h \in H$.

(ii) Choose a 1-cocycle

$$\varphi \in Z^1\big(G, \mathrm{GL}_n(L)\big)\,.$$

For every column vector u of length n with coordinates from L, define the column vector

$$\Psi(u) = \sum_{g\in G} \varphi(g)\cdot {}^g u\,.$$

Let $v \in L^n$ be a row vector such that $v\cdot\Psi(\lambda u) = 0$ for every element $\lambda \in L$. Prove that $v\cdot u = 0$.

Hint. Use the equality

$$v\cdot\Psi(\lambda u) = \sum_{g\in G} (v\cdot\varphi(g)\cdot {}^g u)\;{}^g\lambda$$

together with Artin's lemma applied to the characters $\lambda \mapsto {}^g\lambda$ from L^* to itself.

(iii) Given an $n\times n$ matrix A with entries in L, put

$$\Psi(A) = \sum_{g\in G} \varphi(g)\cdot {}^g A\,.$$

Show that there exists a matrix A such that the matrix $\Psi(A)$ is invertible.

Hint. Use the equality

$$\Psi(u_1,\ldots,u_n) = \big(\Psi(u_1),\ldots,\Psi(u_n)\big)$$

for column vectors $u_1,\ldots,u_n$, and recall that by part (ii) the vectors $\Psi(u)$, $u\in L^n$, generate the vector space L^n.

(iv) Prove that

$${}^g\Psi(A) = \varphi(g)^{-1}\Psi(A)$$

for every $g\in G$.

Hint. Compute ${}^g\Psi(A)$ explicitly, and use the 1-cocycle condition for φ.

Use this together with part (iii) to show that the 1-cocycle φ is trivial.

EXERCISE 2.2.3. **Behavior with respect to field extensions**

Let X be an object in $\mathcal{M}(K)$ and let $K\subset F$ be a finite Galois extension such that there is a tower of Galois extensions

$$K\subset L\subset F\,.$$

Put $N = \mathrm{Gal}(F/L)$ and $H = \mathrm{Gal}(F/K)$, so that one has an exact sequence

$$1\to N\to H\to G\to 1\,.$$

(i) Show that the following diagram is commutative:

$$\begin{array}{ccc} \Phi(X,L) & \longrightarrow & \Phi(X,F) \\ \downarrow & & \downarrow \\ H^1\big(G,\mathrm{Aut}(X_L)\big) & \longrightarrow & H^1\big(H,\mathrm{Aut}(X_F)\big) \end{array}$$

Here the upper horizontal arrow is defined in a natural way, the vertical arrows are maps from Exercise 2.2.1(ii) and (iii), and the lower horizontal arrow is the composition of the pull-back map

$$H^1\big(G,\mathrm{Aut}(X_L)\big)\to H^1\big(H,\mathrm{Aut}(X_L)\big)$$

defined by the group homomorphism $H \twoheadrightarrow G$ with the natural map

$$H^1\big(H, \mathrm{Aut}(X_L)\big) \to H^1\big(H, \mathrm{Aut}(X_F)\big)\,.$$

(ii) Show that the following diagram is commutative:

$$\begin{array}{ccc} \Phi(X,F) & \longrightarrow & \Phi(X_L,F) \\ \downarrow & & \downarrow \\ H^1\big(H,\mathrm{Aut}(X_F)\big) & \longrightarrow & H^1\big(N,\mathrm{Aut}(X_F)\big) \end{array}$$

Here the upper horizontal arrow is given by scalar extension from K to L for forms of X, the vertical arrows are maps from Exercise 2.2.1(ii) and (iii), and the lower horizontal arrow is the pull-back map corresponding to the group homomorphism $N \hookrightarrow H$. Note that this commutative diagram exists even if the extension $K \subset L$ is not Galois.

EXERCISE 2.2.4. **Cohomology with additive coefficients**

(o) What is the group $H^0(G,L)$?

(i) Show that the G-module $K[G]$ is isomorphic to $\iota_*(K)$ (see Definition 1.2.3), where

$$\iota\colon \{e\} \hookrightarrow G$$

is the embedding of the trivial subgroup.

(ii) Prove that

$$H^i\big(G, K[G]\big) = 0$$

for all $i > 0$.

Hint. Apply Shapiro's lemma from Exercise 1.2.5(iii) to the embedding $\{e\} \hookrightarrow G$.

(iii) Prove that

$$H^i(G,L) = 0$$

for all $i > 0$.

Hint. Use the *normal basis theorem*, which states that there is a basis

$$\{e_g \mid g \in G\}$$

of L over K such that $g(e_h) = e_{gh}$ for all $g, h \in G$. Then apply part (ii).

EXERCISE 2.2.5. **First Galois cohomology of certain groups**

(i) Prove that

$$H^1\big(G, \mathrm{SL}_n(L)\big) = \{1\}\,.$$

Hint. Use the exact sequence

$$1 \to \mathrm{SL}_n(L) \to \mathrm{GL}_n(L) \to L^* \to 1$$

together with Hilbert's Theorem 90 and Exercise 1.1.7(ii).

(ii) Show that
$$H^1\big(G, \mathrm{Aff}_n(L)\big) = \{1\},$$
where Aff_n denotes the group of affine automorphisms of the n-dimensional affine space.

Hint. Use the exact sequence
$$1 \to L^n \to \mathrm{Aff}_n(L) \to \mathrm{GL}_n(L) \to 1$$
together with Hilbert's Theorem 90 and Exercise 1.1.7(ii).

(iii) Show that
$$H^1\big(G, \mathrm{Sp}_{2n}(L)\big) = \{1\}$$
if $\mathrm{char}(K) \neq 2$, where Sp_{2n} denotes the symplectic group of $2n \times 2n$ matrices.

Hint. Recall the classification of skew-symmetric forms over K.

(iv) Let S be the set of all complex $n \times n$ matrices M such that
$$\bar{M} = M^T = M^{-1},$$
where $\bar{M}$ denotes the complex conjugate of a matrix M and M^T denotes its transpose. Consider the action of the orthogonal group $\mathrm{O}_n(\mathbb{C})$ on S given by the formula
$$A\colon M \mapsto AM(\bar{A})^{-1}$$
for all $A \in \mathrm{O}_n(\mathbb{C})$ and $M \in S$. How many elements are there in the quotient set $S/\mathrm{O}_n(\mathbb{C})$?

Hint. Use Exercise 1.1.7(v) to show that there is a bijection
$$H^1\big(\mathrm{Gal}(\mathbb{C}/\mathbb{R}), \mathrm{O}_n(\mathbb{C})\big) \cong S/\mathrm{O}_n(\mathbb{C}).$$
Then recall a classification of non-degenerate quadratic forms over $\mathbb{R}$.

DEFINITION 2.2.6. A *Severi–Brauer variety* is a form of a projective space.

EXERCISE 2.2.7. **Forms of $\mathbb{P}^1$**

Prove that the set $\Phi(\mathbb{P}^1)$ is bijective with the set of projective equivalence classes of smooth conics in $\mathbb{P}^2$.

Hint. Consider the embedding of a form X into $\mathbb{P}^2$ given by its very ample anti-canonical sheaf ω_X^{-1}.

EXERCISE 2.2.8. **Severi–Brauer varieties with points**

(o) Show that the set of isomorphism classes of Severi–Brauer varieties that are L-forms of $\mathbb{P}^{n-1}$ is canonically bijective with the set $H^1\big(G, \mathrm{PGL}_n(L)\big)$.

(i) Prove that a Severi–Brauer variety has a point over a field K if and only if it is isomorphic to a projective space over K.

Hint. What is the automorphism group of a projective space with a marked point?

(ii) Let X be a Severi–Brauer variety over a field K. Prove that $X_{K(X)}$ is isomorphic to a projective space over the field $K(X)$ of rational functions on X.

Hint. Construct a $K(X)$-point on $X_{K(X)}$ and apply part (i).

2.3. Cohomology of profinite groups

It is often more convenient to use the absolute Galois group

$$G_K = \mathrm{Gal}(K^{sep}/K)$$

without fixing a finite extension L over K. Since G_K is a profinite group, this leads us to considering cohomology of such groups.

Let $(I, \leqslant)$ be a directed partially ordered set, that is, a (possibly infinite) partially ordered set such that for every pair of elements $i_1, i_2 \in I$ there exists an element $j \in I$ with $i_1 \leqslant j$ and $i_2 \leqslant j$. Suppose that for every $i \in I$ one has a finite group G_i and for every pair of comparable elements $i \leqslant j$ of I one has a group homomorphism

$$\varphi_{ji} \colon G_j \to G_i$$

such that $\varphi_{ii} = \mathrm{id}_{G_i}$ and

$$\varphi_{ji} \circ \varphi_{kj} = \varphi_{ki}$$

for all $i \leqslant j \leqslant k$. The collection $\{G_i\}$, $i \in I$, is called an *inverse system of finite groups*.

Recall that on a (possibly infinite) product of discrete finite groups $\prod_{i\in I} G_i$ one has the *Tychonoff topology*, also known as the *product topology*: the neighborhood system of unity is given by subsets of the form

$$U_S = \prod_{i\in S} \{e\} \times \prod_{i\in G\setminus S} G_i \,,$$

where $S \subset I$ is a finite subset. The *inverse limit* $\varprojlim G_i$ consists of collections

$$(g_i)_{i\in I} \in \prod_{i\in I} G_i$$

such that $\varphi_{ji}(g_j) = g_i$ for all $i \leqslant j$. We will consider the topology on the set $\varprojlim G_i$ restricted from Tychonoff topology on $\prod_{i\in I} G_i$. One can easily check that the inverse limit of an inverse system of groups is again a group.

DEFINITION 2.3.1. A *profinite group* is a topological group isomorphic to an inverse limit $\varprojlim G_i$ for some inverse system of finite groups $\{G_i\}$, $i \in I$.

EXAMPLE 2.3.2. The following topological groups are profinite:

(i) the additive group of p-adic numbers $\mathbb{Z}_p = \varprojlim \mathbb{Z}/p^n\mathbb{Z}$;

(ii) the group $\widehat{\mathbb{Z}} = \varprojlim \mathbb{Z}/n\mathbb{Z}$;

(iii) the *profinite completion* of an arbitrary group G, which is defined as

$$\widehat{G} = \varprojlim G/N \,,$$

where the limit is taken over all normal subgroups N of finite index in G;

(iii) the *pro-p-completion* of an arbitrary group G, which is defined as

$$\widehat{G}_p = \varprojlim G/N \,,$$

where the limit is taken over all normal subgroups N whose index in G is a power of p.

EXERCISE 2.3.3. **Profinite groups**

(o) Show that the projections from the product $\prod_i G_i$ to G_j give rise to homomorphisms
$$\pi_j\colon \varprojlim G_i \to G_j\,.$$
Prove that $\varprojlim G_i$ has the following universal property: for every group H with homomorphisms
$$\xi_i\colon H \to G_i$$
such that $\varphi_{ji} \circ \xi_j = \xi_i$ for all $i \leqslant j$, there exists a unique homomorphism
$$\xi\colon H \to \varprojlim G_i$$
such that $\pi_j \circ \xi = \xi_j$ for all $j \in I$.

(i) Show that a profinite group is compact.
Hint. First prove that the product $\prod_{i\in I} G_i$ is compact with respect to the Tychonoff topology. Then show that $\varprojlim G_i$ is a closed subset in $\prod_{i\in I} G_i$.

(ii) Show that the topology on $\varprojlim G_i$ is the weakest topology with respect to which all natural projections
$$\pi_j\colon \varprojlim G_i \to G_j$$
are continuous. Show that the subgroups
$$\operatorname{Ker}(\pi_j) \subset \varprojlim G_i$$
are both open and closed.

(iii) Prove that every profinite group is isomorphic to its profinite completion.

(iv) Prove that every open subgroup of a profinite group has finite index.
Hint. Show that the set of cosets is both discrete and compact with respect to the quotient topology.

EXERCISE 2.3.4. **Profinite Galois theory**

Let $L \subset K$ be a Galois extension, that is, a separable normal (but possibly infinite) algebraic extension of fields.

(i) Construct an isomorphism
$$\operatorname{Gal}(L/K) \cong \varprojlim \operatorname{Gal}(L_i/K)\,,$$
where the limit is taken over all subfields L_i of L that are finite Galois extensions of K. Therefore, the Galois group $\operatorname{Gal}(L/K)$ is profinite.

(ii) Show that a base of open neighborhoods of unity in $\operatorname{Gal}(L/K)$ is given by stabilizers of elements of L.

(iii) Prove the *fundamental theorem of Galois theory*: there is a natural bijection between the set of all closed subgroups $H \subset \operatorname{Gal}(L/K)$ and the set of all intermediate extensions $K \subset E \subset L$ given by the usual formula
$$E = L^H, \quad H = \operatorname{Gal}(L/E)\,.$$
Moreover, a subgroup H is normal in $\operatorname{Gal}(L/K)$ if and only if the corresponding field E is normal over K.
Hint. Given a subfield E, to show that the subgroup $\operatorname{Gal}(L/E)$ is closed use the fact that
$$\operatorname{Gal}(L/E) = \bigcap_i \operatorname{Gal}(L/E_i)\,,$$

where E_i varies in the set of all subfields $E_i \subset E$ that are finite over K. Also keep in mind that $\mathrm{Gal}(L/E_i)$ is a closed subgroup in $\mathrm{Gal}(L/K)$: this can be proved by applying Exercise 2.3.3(ii) to homomorphisms

$$\mathrm{Gal}(L/K) \to \mathrm{Gal}(F_i/K),$$

where F_i denotes the normal closure of E_i over K in L. For a given closed subgroup

$$H \subset \mathrm{Gal}(L/K),$$

show that H is also a closed subgroup in the group $\mathrm{Gal}(L/E)$, where $E = L^H$. Applying the usual Galois theory, show that for every normal subgroup of finite index

$$N \subset \mathrm{Gal}(L/E),$$

the induced map

$$H \to \mathrm{Gal}(L/E)/N$$

is surjective. Finally, use the fact that H is closed in $\mathrm{Gal}(L/E)$ to deduce that $H = \mathrm{Gal}(L/E)$.

For instance, one has an isomorphism

$$\mathrm{Gal}(\bar{\mathbb{F}}_p/\mathbb{F}_p) \cong \widehat{\mathbb{Z}}$$

for every finite field $\mathbb{F}_p$. This profinite group is topologically generated by the Frobenius map

$$\mathrm{Fr}\colon \bar{\mathbb{F}}_p \to \bar{\mathbb{F}}_p, \quad x \mapsto x^p.$$

EXERCISE 2.3.5. **Discrete modules**

(i) Let G be a profinite group, and let M be a G-module with discrete topology. Show that the action map

$$G \times M \to M$$

is continuous if and only if stabilizers of all elements of M are open in G (recall that by Exercise 2.3.3(iv) an open subgroup of a profinite group has finite index). Such G-modules are called *discrete*.

(ii) Show that for every discrete G-module M one has

$$M = \bigcup_U M^U,$$

where U varies in the set of all open normal subgroups in G.

Let $(I, \leqslant)$ be as before, a directed partially ordered set. Let $\{X_i\}_{i\in I}$ be a *direct system of sets*; that is, for every pair of comparable elements $i \leqslant j$ one has a map

$$\psi_{ij}\colon X_i \to X_j$$

such that $\psi_{ii} = \mathrm{id}_{X_i}$ and $\psi_{jk} \circ \psi_{ij} = \psi_{ik}$ for all $i \leqslant j \leqslant k$. The *direct limit* $\varinjlim X_i$ is defined as the quotient set

$$\Big(\coprod_{i\in I} X_i\Big)/\big(\varphi_{ij}(x_i) \sim x_j\big).$$

Alternatively, one can define the direct limit via a universal property; cf. Exercise 2.3.3(o). It is straightforward to check that a direct limit of a direct system of abelian groups is again an abelian group.

DEFINITION 2.3.6. Let M be a discrete G-module, and let $U_2 \subset U_1$ be open normal subgroups in G. Consider the inflation map

$$H^i\big(G/U_1, M^{U_1}\big) \to H^i\big(G/U_2, M^{U_2}\big), \quad i \geqslant 0\,,$$

that is, the composition of the pull-back map

$$H^i\big(G/U_1, M^{U_1}\big) \to H^i\big(G/U_2, M^{U_1}\big)$$

and the natural map

$$H^i\big(G/U_2, M^{U_1}\big) \to H^i\big(G/U_2, M^{U_2}\big)\,.$$

This gives a direct system of abelian groups. The *cohomology groups of the profinite group G with coefficients in the discrete module M* are defined as

$$H^i(G, M) = \varinjlim H^i\big(G/U, M^U\big)\,,$$

where the direct limit is taken over all normal subgroups $U \subset G$. Similarly one defines cohomology groups $H^0(G, \Gamma)$ and $H^1(G, \Gamma)$ for a non-abelian discrete G-module Γ (cf. Exercise 1.1.7).

Note that cohomology of a profinite group G with coefficients in a discrete G-module M can differ from cohomology of G considered as an abstract group with coefficients in M (see Definition 1.1.2). Still, we will use the same notation for both of these types of cohomology for simplicity. In the rest of the book, given a profinite group G and a discrete G-module M, by $H^i(G, M)$ we will always mean the cohomology groups in the sense of Definition 2.3.6.

The following exercise shows that cohomology groups of profinite groups have the same basic properties as those of finite groups.

EXERCISE 2.3.7. **Cohomology of profinite groups**
Let G be a profinite group, and let M be a discrete G-module.

(o) Show that $H^0(G, M) \cong M^G$.

(i) Consider a complex $\mathrm{Map}_c(G^\bullet, M)$ whose terms are groups of continuous maps from G^i to M, $i \geqslant 0$, with the differential defined in the same way as in the complex (1.1) from Exercise 1.1.3(v). Prove that

$$H^i(G, M) \cong H^i\big(\mathrm{Map}_c(G^\bullet, M)\big)\,.$$

Hint. Show that

$$\mathrm{Map}_c\left(G^\bullet, M\right) \cong \varinjlim \mathrm{Map}\left((G/U)^\bullet, M^U\right),$$

where the direct limit is taken over all open normal subgroups $U \subset G$. Then use the fact that taking cohomology groups commutes with taking direct limits.

(ii) Let M be a trivial G-module. Show that

$$H^1(G, M) \cong \mathrm{Hom}(G, M)\,,$$

where $\mathrm{Hom}(G, M)$ denotes the group of continuous homomorphisms, that is, homomorphisms with finite image.

(iii) Show that $H^1(G, \mathbb{Z}) = 0$ and there is a canonical isomorphism

$$H^2(G, \mathbb{Z}) \cong \mathrm{Hom}(G, \mathbb{Q}/\mathbb{Z})\,,$$

where $\mathbb{Z}$ and $\mathbb{Q}/\mathbb{Z}$ are considered as trivial G-modules.

Hint. Apply Exercise 1.3.3(ii).

(iv) Show that $H^i(G, M)$ are torsion groups for all $i > 0$.

Hint. Use Exercise 1.3.1(i).

(v) Show that a short exact sequence of discrete G-modules

$$0 \to M' \to M \to M'' \to 0$$

gives rise to a long exact sequence of cohomology groups.

(vi) Prove that

$$H^i(G, M) \cong \mathrm{Ext}^i_G(\mathbb{Z}, M)\,,$$

where Ext is taken in the abelian category of discrete G-modules.

(vii) Let G be a profinite group, H a finite group, and $f\colon G \twoheadrightarrow H$ a surjective homomorphism with kernel I. Let M be a discrete G-module. (Check that I is a profinite group and M is a discrete I-module!) Show that there is an exact sequence

$$0 \to H^1(H, M^I) \to H^1(G, M) \to H^1(I, M)^H \to H^2(H, M^I) \to H^2(G, M)\,.$$

Hint. Apply Exercise 1.2.6(iv) and take a limit.

Actually, the same assertion holds when one assumes only that the profinite subgroup $I \subset G$ is closed instead of finiteness of the group H.

Exercise 2.3.8. **First cohomology of the group $\widehat{\mathbb{Z}}$**

Let M be a discrete module over the group $\widehat{\mathbb{Z}}$. Suppose that every element of M has finite order (as an element of the abelian group M). Show that

$$H^1(\widehat{\mathbb{Z}}, M) \cong M/(s-1)M\,,$$

where $s \in \widehat{\mathbb{Z}}$ is an arbitrary topological generator of the group $\widehat{\mathbb{Z}}$.

Hint. Prove that for every element $m \in M$ there exists a positive integer n such that

$$\sum_{i=0}^{n-1} s^i(m) = 0\,.$$

Then recall the explicit description of cohomology of finite cyclic groups obtained in Exercise 1.1.5(i), and take a direct limit as in Definition 2.3.6.

Exercise 2.3.9. **Finiteness of first cohomology groups**

Let G be a profinite group, and let M be a discrete G-module. Suppose that M is finitely generated as an abelian group.

(i) Show that the kernel I of the action of the group G on the module M has finite index in G.

Hint. Choose a set of generators

$$m_1, \ldots, m_r \in M$$

of M as an abelian group, consider the intersection $I \subset G$ of the stabilizers of the elements m_i in G, and apply Exercise 2.3.3(iv).

(ii) Suppose that M is free as an abelian group. Check that

$$H^1(I, M) = 0\,.$$

Hint. Use Exercise 2.3.7(ii).

(iii) Suppose that M is free as an abelian group. Put $H = G/I$. Then M can be considered as a discrete H-module. Prove that

$$H^1(G,M) \cong H^1(H,M)\,.$$

Hint. Apply part (ii) and Exercise 2.3.7(vii).

In particular, the group $H^1(G,M)$ is finite by Exercise 1.3.1(ii).

(iv) What changes if one drops the assumption that M is free as an abelian group in part (iii)?

Hint. In this case the assertion of part (iii) does not hold even for a finite group G.

EXERCISE 2.3.10. **Cohomology of profinite groups with coefficients in a permutation module**

Let G be a profinite group, and let M be a discrete G-module.

(i) Suppose that M is a permutation G-module. Let B be a basis in M such that G acts by permutations on B. Denote by $\mathbb{O}$ the set of all G-orbits in B. Prove that $H^1(G,M)=0$ and

$$H^2(G,M) \cong \bigoplus_{O\in\mathbb{O}} \mathrm{Hom}\left(\mathrm{Stab}_G(x_O), \mathbb{Q}/\mathbb{Z}\right),$$

where x_O is an (arbitrary) element in an orbit O, the subgroup $\mathrm{Stab}_G(x_O)$ is the stabilizer of x_O in the group G, and $G/\mathrm{Stab}_G(x_O)$ denotes the set of cosets of this subgroup.

Hint. Apply Exercise 1.4.2(ii).

(ii) Suppose that M is a stably permutation G-module. Prove that

$$H^1(G,M)=0\,.$$

Hint. Apply Exercise 1.4.5(i).

2.4. Cohomology of the absolute Galois group

In the rest of this chapter we will keep the notation G_K for the absolute Galois group of the field K, so that $G_K = \mathrm{Gal}(K^{sep}/K)$, where K^{sep} is the separable closure of K.

EXERCISE 2.4.1. **First cohomology and forms**

(i) Let $\mathcal{M}$ be a category fibered over fields that satisfies descent, and let X be an object in $\mathcal{M}(K)$. Show that $\mathrm{Aut}(X_{K^{sep}})$ is a discrete G_K-module.
(ii) Show that there is a canonical bijection

$$\Phi(X) \cong H^1\big(G_K, \mathrm{Aut}(X_{K^{sep}})\big)\,.$$

Hint. Use Exercises 2.2.1(iii) and 2.2.3.

EXERCISE 2.4.2. **Kummer theory**

Let n be a positive integer coprime to the characteristic of the field K.

(o) Show that

$$H^1(G_K, \mathrm{GL}_d(K^{sep})) = \{1\}$$

for any $d \geqslant 1$.

Hint. Apply Hilbert's Theorem 90 and take a limit.

In particular, one has $H^1(G_K, (K^{sep})^*) = \{1\}$.

(i) Show that there is a short exact sequence of discrete G_K-modules

$$1 \to \mu_n \longrightarrow (K^{sep})^* \xrightarrow{n} (K^{sep})^* \to 1\,,$$

where μ_n denotes the group of nth roots of unity in the field K^{sep} and the second homomorphism is raising to the nth power.

(ii) Suppose that the field K contains all elements from μ_n. Construct an isomorphism of abelian groups

$$\mathrm{Hom}(G_K, \mu_n) \cong K^*/(K^*)^n\,.$$

Hint. Use the long exact sequence of cohomology groups associated with the short exact sequence of G_K-modules from part (i).

(iii) Under the assumptions of part (ii), prove that every finite Galois extension $K \subset L$ with a cyclic Galois group whose order divides n has the form

$$L = K(\sqrt[n]{a})$$

for some element $a \in K^*$.

Hint. For every element $a \in K^*$ one constructs a map

$$G_K \to \mu_n, \quad g \mapsto g(\sqrt[n]{a})/\sqrt[n]{a}\,.$$

Compare this map with the image of the element

$$a \in K^* \cong H^0(G_K, (K^{sep})^*)$$

under the coboundary map in the long exact sequence associated with the short exact sequence from part (i); see Exercise 1.1.6(i).

EXERCISE 2.4.3. **Artin–Schreier theory**
Let K be a field of characteristic $p > 0$.

(o) Show that

$$H^i(G_K, K^{sep}) = 0$$

for every $i > 0$.

Hint. Apply Exercise 2.2.4(iii) and take a limit.

Actually, this assertion holds for fields of zero characteristic as well.

(i) Show that there is a short exact sequence of discrete G_K-modules

$$0 \to \mathbb{Z}/p\mathbb{Z} \longrightarrow K^{sep} \xrightarrow{\mathrm{Fr}-1} K^{sep} \to 0\,,$$

where Fr denotes the Frobenius morphism $\mathrm{Fr}(x) = x^p$.

(ii) Construct an isomorphism of abelian groups

$$\mathrm{Hom}(G_K, \mathbb{Z}/p\mathbb{Z}) \cong K/(\mathrm{Fr}-1)K\,.$$

Hint. Use the long exact sequence of cohomology groups associated with the short exact sequence of G_K-modules from part (i).

(iii) Prove that every finite Galois extension $K \subset L$ with Galois group $\mathbb{Z}/p\mathbb{Z}$ has the form $L = K(x)$, where an element $x \in K^{sep}$ satisfies the condition

$$x^p - x = a$$

for some $a \in K$.

Hint. For every element $a \in K$ one constructs a map

$$G_K \to \mathbb{Z}/p\mathbb{Z}, \quad g \mapsto g(x) - x\,,$$

where $x \in K^{sep}$ satisfies the condition $x^p - x = a$. Compare this map with the image of the element

$$a \in K \cong H^0(G_K, K^{sep})$$

under the coboundary map in the long exact sequence associated with the short exact sequence from part (i); see Exercise 1.1.6(i).

2.5. Picard group as a stably permutation module

Now we will produce examples of discrete stably permutation Galois modules that arise from algebraic geometry. Let X be a smooth algebraic variety over the field K. The group of divisors $\mathrm{Div}(X_{K^{sep}})$ and the Picard group $\mathrm{Pic}(X_{K^{sep}})$ are discrete G_K-modules.

EXERCISE 2.5.1. **Cohomology of the group of divisors**

(i) Show that the group $\mathrm{Div}(X_{K^{sep}})$ is a permutation G_K-module.

(ii) Let $K \subset F$ be a (possibly infinite) extension of fields, and let K_F be the separable closure of the field K in the field F. Suppose that the degree $n = [K_F : K]$ is finite. Show that there exists an isomorphism of K^{sep}-algebras

$$K^{sep} \otimes_K F \cong \prod_{i=1}^{n} F_i\,,$$

where F_i are fields containing K^{sep}. Moreover, the Galois group G_K acts transitively on the set of the fields $\{F_i\}$. Let F_{i_0} be one of these fields. Show that the stabilizer of F_{i_0} corresponds to a subfield in K^{sep} which coincides with the image of K_F under the composition

$$K_F \subset F \subset F_{i_0}\,,$$

where the second embedding is given by the projection $K^{sep} \otimes_K F \to F_{i_0}$. Don't forget to check that this image is contained in $K^{sep} \subset F_{i_0}$.

Hint. Use the isomorphism

$$K^{sep} \otimes_K F \cong (K^{sep} \otimes_K K_F) \otimes_{K_F} F\,,$$

the isomorphism

$$K^{sep} \otimes_K K_F \cong \prod_{i=1}^{n} K^{sep}\,,$$

and the fact that $K^{sep} \otimes_{K_F} F$ is an integral domain since K_F is separably closed in F; see [**Jac75**, Theorem IV.21(2)].

(iii) Prove that there are canonical isomorphisms

$$H^0\big(G_K, \mathrm{Div}(X_{K^{sep}})\big) \cong \mathrm{Div}(X), \quad H^1\big(G_K, \mathrm{Div}(X_{K^{sep}})\big) = 0,$$

and

$$H^2\big(G_K, \mathrm{Div}(X_{K^{sep}})\big) \cong \bigoplus_{D\subset X} \mathrm{Hom}\left(G_{K_D}, \mathbb{Q}/\mathbb{Z}\right),$$

where D varies in the set of all reduced irreducible divisors on X and K_D is the separable closure of the field K in the field $K(D)$ of rational functions on D.

Hint. A choice of an irreducible component E of the divisor $D_{K^{sep}} \subset X_{K^{sep}}$ over K^{sep} provides an embedding

$$K(D) \subset K^{sep}(E),$$

which in turn defines an embedding $K_D \subset K^{sep}$. It follows from part (ii) that the corresponding Galois subgroup $G_{K_D} \subset G_K$ is the stabilizer of the irreducible component E. Now one can apply part (i) and Exercise 2.3.7.

In the following exercise we will assume that the characteristic of the field K equals zero. We will need this assumption to use Hironaka's desingularization theorem. Also, in this case one has $K^{sep} = \bar{K}$.

EXERCISE 2.5.2. **Stably permutation modules and the Picard group**

(i) Let $\pi\colon \widetilde{X} \to X$ be a proper birational morphism of smooth varieties over K. Use the direct image and inverse image maps

$$\pi_*\colon \mathrm{Pic}(\widetilde{X}_{\bar{K}}) \to \mathrm{Pic}(X_{\bar{K}}), \quad \pi^*\colon \mathrm{Pic}(X_{\bar{K}}) \to \mathrm{Pic}(\widetilde{X}_{\bar{K}})$$

to construct an isomorphism of G_K-modules

$$\mathrm{Pic}(\widetilde{X}_{\bar{K}}) \cong \mathrm{Pic}(X_{\bar{K}}) \oplus N,$$

where $N = \mathrm{Ker}(\pi_*)$.

Hint. Observe that

$$\pi_* \circ \pi^*\colon \mathrm{Pic}(X_{\bar{K}}) \to \mathrm{Pic}(X_{\bar{K}})$$

is the identity map. This follows from the fact that the set of indeterminacy points of the rational map

$$\pi^{-1}\colon X \dashrightarrow \widetilde{X}$$

has codimension at least two.

(ii) Let Σ denote the set of all exceptional divisors of the morphism π, that is, irreducible divisors $D \subset \widetilde{X}$ such that

$$\dim\big(\pi(D)\big) < \dim(D).$$

Show that there is a canonical isomorphism of G_K-modules

$$\mathbb{Z}[\Sigma] \cong N.$$

Hint. First construct a surjective morphism of G_K-modules

$$\mathbb{Z}[\Sigma] \to N.$$

Then prove that it is injective, again using the fact that the rational map π^{-1} is well-defined outside of a subset of codimension two.

(iii) Use part (ii) to show that N is a permutation G_K-module.

(iv) Suppose that the variety X is rational over K and projective. Prove that $\mathrm{Pic}(X_{\bar{K}})$ is a stably permutation G_K-module.

Hint. By Hironaka's desingularization theorem there exists a smooth projective variety $\widetilde{X}$ together with proper birational morphisms $\widetilde{X} \to \mathbb{P}^n$ and $\widetilde{X} \to X$, where $n = \dim(X)$. Now one can apply parts (i) and (iii).

(v) Suppose that the variety X is stably rational over K (see Definition 4.3.3 below) and projective. Prove that $\mathrm{Pic}(X_{\bar{K}})$ is a stably permutation G_K-module.

Hint. Use the isomorphism of G_K-modules

$$\mathrm{Pic}\left((X \times \mathbb{P}^n)_{\bar{K}}\right) \cong \mathrm{Pic}\left(X_{\bar{K}}\right) \oplus \mathbb{Z}$$

and part (iv).

2.6. Torsors

Let U be a variety defined over a field K, and let Γ be a (reduced) algebraic group over K. An *action* of the group Γ on U is a morphism

$$m\colon \Gamma \times U \to U$$

satisfying the usual associativity condition. Equivalently, for every K-scheme S the morphism m defines an action of the group $\Gamma(S)$ on the set $U(S)$.

DEFINITION 2.6.1. An action of Γ on U is *free and transitive* if $U \neq \varnothing$ (that is, U has a point over $\bar{K}$) and the morphism

$$(m, \mathrm{pr}_U)\colon \Gamma \times U \to U \times U$$

is an isomorphism. Equivalently, for every K-scheme S and every point $x \in U(S)$, the map

$$g \mapsto m(g, x)$$

defines a bijection between $\Gamma(S)$ and $U(S)$. Note that it is enough to check the latter condition for the scheme $S = \mathrm{Spec}(K^{sep})$. If the action of Γ on U is free and transitive, we say that U is a *torsor* under Γ (or a Γ-*torsor*, or a *principal homogeneous space* over Γ). If $U(K) \neq \varnothing$, then U and Γ are isomorphic as varieties over K, and U is said to be a *trivial* Γ-torsor. The notion of isomorphism for torsors is defined in a natural way.

EXERCISE 2.6.2. **Torsors and first Galois cohomology**

(i) Prove that the set of isomorphism classes of Γ-torsors over the field K is bijective with the set $H^1\big(G_K, \Gamma(K^{sep})\big)$.

Hint. Use the fact that an arbitrary torsor is a form of a trivial torsor. Also show that the automorphism group of the trivial torsor over K^{sep} is isomorphic to the group $\Gamma(K^{sep})$, and apply Exercise 2.2.1(iii).

(ii) Let Γ be one of the groups GL_n, SL_n, Aff_n, and Sp_{2n} (in the latter case we will also assume that $\mathrm{char}(K) \neq 2$). Prove that all Γ-torsors are trivial.

Hint. Use part (i) together with Hilbert's Theorem 90 and Exercise 2.2.5.

A simple example of a torsor under the group $\mu_2 \cong \mathbb{Z}/2\mathbb{Z}$ is the subvariety in $\mathbb{A}^1$ given by the equation

$$\{x^2 = a\}$$

for some $a \in K^*$. Definition 2.6.1 easily implies that this torsor is trivial if and only if $a \in (K^*)^2$. By Exercise 2.6.2(i), this corresponds to isomorphisms

$$H^1(G_K, \mu_2) \cong \mathrm{Hom}(G_K, \mu_2) \cong K^*/(K^*)^2\,;$$

cf. Exercise 2.4.2(ii).

DEFINITION 2.6.3. Let $\phi\colon V \to X$ be a morphism of varieties defined over a field k. Suppose that ϕ is flat (see [**Har77**, III.9]) and surjective on $\bar{k}$-points. Suppose that a (reduced) algebraic group Γ acts on the variety V so that the action is fiberwise with respect to ϕ and is free and transitive on its fibers. In this case we say that V is a Γ-*torsor over* X. (Sometimes one says that V is a *torsor under* Γ, if there is no need to mention the base and no confusion is likely to arise.) A notion of isomorphism for torsors over X is defined in a natural way.

EXERCISE 2.6.4. **Birational properties of torsors**
Let X be an irreducible variety over the field k, and let $\phi\colon V \to X$ be a Γ-torsor. Consider the corresponding Γ-torsor U over the function field $K = k(X)$.

(o) Check that the torsor U is trivial if and only if the morphism ϕ has a rational section.
(i) Using part (o), check that if the torsor U is trivial, then the varieties V and $X \times \Gamma$ are birational (assume for simplicity that Γ is geometrically irreducible—see §3.3 below for a definition; this implies that the variety $X \times \Gamma$ is irreducible).
(ii) Suppose that $H^1\big(G_K, \Gamma(K^{sep})\big) = \{1\}$; for instance, Γ is one of the groups listed in Exercise 2.6.2(ii). Prove that the varieties V and $X \times \Gamma$ are birational.

Hint. Apply part (i) together with Exercise 2.6.2(i).

2.7. Cohomology of the inverse limit

Let $\{C_i^\bullet\}$ be an inverse system of complexes of abelian groups, where i varies in the set of positive integers with natural order (see §2.3). Thus for every pair of positive integers $i \leqslant j$ we have a morphism of complexes

$$\varphi_{ji}\colon C_j^\bullet \to C_i^\bullet\,,$$

and the compatibility conditions are satisfied. Then for every integer p there is a well-defined abelian group

$$C^p = \varprojlim C_i^p\,,$$

and these groups form a complex

$$C^\bullet = \varprojlim C_i^\bullet\,.$$

For every integer p there is a canonical map

$$H^p(C^\bullet) \to \varprojlim H^p(C_i^\bullet)\,.$$

EXERCISE 2.7.1. **Cohomology of the inverse limit of complexes**

(i) Let

$$\varphi\colon E^\bullet \to D^\bullet$$

be a surjective morphism of complexes. Suppose that for a cocycle $\alpha \in D^p$ there is a class $c \in H^p(E^\bullet)$ such that one has $\varphi(c) = [\alpha]$ in $H^p(D^\bullet)$. Show that there exists a cocycle $\beta \in E^p$ such that its class in $H^p(E^\bullet)$ equals c and one has $\varphi(\beta) = \alpha$ in D^p.

Hint. First consider an arbitrary cocycle $\beta' \in E^p$ such that $[\beta'] = c$. There exists an element $\zeta \in D^{p-1}$ with the property that

$$\alpha = \varphi(\beta') + d(\zeta)\,.$$

Now put

$$\beta = \beta' + d(\xi)\,,$$

where $\xi \in E^{p-1}$ is an element such that $\varphi(\xi) = \zeta$.

(ii) Suppose that all homomorphisms φ_{ji} are surjective. Prove that for every integer p the canonical map

$$H^p(C^\bullet) \to \varprojlim H^p(C_i^\bullet)$$

is an isomorphism; in other words, taking cohomology groups commutes with taking the inverse limit in this case.

Hint. Apply part (i).

EXERCISE 2.7.2. **Cohomology of an inverse limit of modules**

Let G be an arbitrary group and M a G-module. Suppose that there is a decreasing filtration of M by G-submodules,

$$M = M^0 \supset M^1 \supset \ldots \supset M^i \supset M^{i+1} \supset \ldots.$$

Suppose also that the natural map

$$M \to \varprojlim M/M^i$$

is an isomorphism.

(i) Show that the natural map

$$H^p(G, M) \to \varprojlim H^p(G, M/M^i)$$

is an isomorphism for every $p \geqslant 0$.

Hint. Apply Exercise 2.7.1(ii) to the inverse limit of complexes

$$C_i^\bullet = \mathrm{Map}(G^{\times\bullet}, M/M^i)\,;$$

see Exercise 1.1.3.

(ii) Suppose that for some $p \geqslant 0$ and for every $i \geqslant 1$ one has

$$H^p(G, M^i/M^{i+1}) = H^{p+1}(G, M^i/M^{i+1}) = 0\,.$$

Put $\overline{M} = M/M^1$. Prove that the natural map

$$H^p(G, M) \to H^p(G, \overline{M})$$

is an isomorphism.

Hint. Prove by induction that for every $i \geqslant 1$ the map

$$H^p(G, M/M^i) \to H^p(G, \overline{M})$$

is an isomorphism, and apply part (i).

EXERCISE 2.7.3. **Cohomology of an inverse limit of non-abelian modules**

Let G be a group acting by automorphisms on a (possibly non-abelian) group Γ. Suppose that there is a decreasing filtration of Γ by normal G-invariant subgroups,

$$\Gamma \supset \Gamma^1 \supset \ldots \supset \Gamma^i \supset \Gamma^{i+1} \supset \ldots .$$

Suppose that the natural map

$$\Gamma \to \varprojlim \Gamma/\Gamma^i$$

is an isomorphism.

(i) Show that the natural map

$$H^1(G, \Gamma) \to \varprojlim H^1(G, \Gamma/\Gamma^i)$$

is bijective.

Hint. Modify the arguments from Exercises 2.7.2(i) and 2.7.1.

(ii) Under the assumptions of part (i), suppose that for every $i \geqslant 1$ the quotient group Γ^i/Γ^{i+1} is abelian and

$$H^1(G, \Gamma^i/\Gamma^{i+1}) = H^2(G, \Gamma^i/\Gamma^{i+1}) = 0\,.$$

Put $\overline{\Gamma} = \Gamma/\Gamma^1$. Prove that the natural map

$$H^1(G, \Gamma) \to H^1(G, \overline{\Gamma})$$

is bijective.

Hint. Do the same as in Exercise 2.7.2(ii).

2.8. Further reading

Exercises 1.4.2(v) and 2.5.2 imply that for a stably rational smooth projective variety X defined over a field K of characteristic zero, the group $H^1\big(G, \mathrm{Pic}(X_{\bar{K}})\big)$ vanishes. Moreover, using a similar argument to that in the solution of Exercise 1.4.2(iv), one can show that this group is a stable birational invariant of smooth projective varieties over K. This invariant has been used in various contexts; see for instance [**Man86**, § IV]. For the geometric version of this invariant we refer the reader to [**BP13**], [**Pro15**], and [**Shi17**].

Part II

Brauer group

CHAPTER 3

Brauer Group of a Field

3.1. Definition and basic properties

We start by recalling several notions and facts concerning associative algebras (see [**Ser79**, Chapter X], [**Bou58**, Chapter VIII], or [**GS06**, Chapter 2] for details). Let K be a field and A an associative unital algebra over K. The algebra A is said to be *central over* K if the center of A coincides with K. The algebra A is *simple* if it does not contain two-sided ideals except for the zero ideal and A itself. The class of finite-dimensional central simple algebras is closed with respect to the tensor product; see [**Bou58**, 1.2, 7.4]. Let $\mathrm{M}_n(K)$ denote the algebra of $n \times n$ matrices with entries from the field K. Note that $\mathrm{M}_n(K)$ is a central simple algebra over K.

Let A be a finite-dimensional central simple algebra over K. *Wedderburn's theorem* (see [**Bou58**, 5.4]) states that there is a finite-dimensional simple division algebra D over K such that $A \cong \mathrm{M}_m(D)$ for some positive integer m. Note that

$$\mathrm{M}_m(D) \cong D \otimes_K \mathrm{M}_m(K)\,.$$

Moreover, if for some finite-dimensional division algebras D_1 and D_2 over K and a positive integer m one has an isomorphism

$$\mathrm{M}_m(D_1) \cong \mathrm{M}_m(D_2)\,,$$

then $D_1 \cong D_2$.

There are no finite-dimensional central division algebras over a separably closed field except for the field itself; see [**Bou58**, 10.5]. Hence every finite-dimensional central simple algebra over a separably closed field is isomorphic to a matrix algebra. On the other hand, a finite-dimensional central simple algebra remains central and simple after scalar extension; see [**Bou58**, 7.4]. Thus the dimension of a finite-dimensional central simple algebra over an arbitrary field is a square.

We will say that two finite-dimensional central simple algebras A and B are equivalent (and will write $A \sim B$ in this case) if there is a finite-dimensional central division algebra D over K and positive integers p and q such that $A \cong \mathrm{M}_p(D)$ and $B \cong \mathrm{M}_q(D)$. In other words, our equivalence relation is the symmetric transitive closure of the relation

$$A \sim A \otimes_K \mathrm{M}_m(K)\,.$$

Every equivalence class contains a unique central division algebra. Also, if $A \sim B$ and $\dim_K(A) = \dim_K(B)$, then $A \cong B$. By $\mathrm{Br}(K)$ we denote the set of equivalence classes of finite-dimensional central simple algebras over K. Given an algebra A, we denote by $[A]$ its class in $\mathrm{Br}(K)$. We consider $\mathrm{Br}(K)$ as an abelian semigroup with its operation given by the tensor product of algebras. Note that one has $[A] = 0$ in this semigroup if and only if A is a matrix algebra over the field K.

For an arbitrary algebra A we denote by A^{op} the *opposite algebra*, that is, the underlying vector space of A with multiplication performed in the opposite order to the multiplication in A. Note that there is an isomorphism of algebras

$$A \otimes_K A^{op} \cong \mathrm{End}_K(A) \cong \mathrm{M}_n(K),$$

where $n = \dim_K(A)$, given by the formula

$$a \otimes b \mapsto \{x \mapsto axb\}, \quad a, x \in A,\ b \in A^{op}.$$

Thus every element in $\mathrm{Br}(K)$ has an inverse, and $\mathrm{Br}(K)$ is an abelian group. This group is an important arithmetic invariant of the field.

DEFINITION 3.1.1. The group $\mathrm{Br}(K)$ is called the *Brauer group* of the field K.

A field extension $K \subset L$ defines a homomorphism of Brauer groups

$$\mathrm{Br}(K) \to \mathrm{Br}(L)$$

given by scalar extension for algebras $A \mapsto L \otimes_K A$. Put

$$\mathrm{Br}(L/K) = \mathrm{Ker}\left(\mathrm{Br}(K) \to \mathrm{Br}(L)\right).$$

In particular, one has

$$\mathrm{Br}(K) = \bigcup_L \mathrm{Br}(L/K),$$

where L varies in the set of all finite Galois extensions of the field K, because it follows from the above that $\mathrm{Br}(K^{sep}) = 0$.

If D is a central division algebra of dimension n^2 over K, then every maximal subfield $L \subset D$ has dimension n over K; see [**Bou58**, 10.3]. Moreover, the element $[D] \in \mathrm{Br}(K)$ corresponding to D is contained in the subgroup

$$\mathrm{Br}(L/K) \subset \mathrm{Br}(K);$$

see [**Bou58**, 10.5].

EXERCISE 3.1.2. **Cohomological definition of the Brauer group**
Let $K \subset L$ be a finite Galois extension with Galois group G.

(i) Construct a bijection between the set $\Phi\big(\mathrm{M}_n(K), L\big)$ of isomorphism classes of L-forms of the matrix algebra $\mathrm{M}_n(K)$ and the set $H^1\big(G, \mathrm{PGL}_n(L)\big)$.

Hint. Use the fact that all automorphisms of the matrix algebra are conjugations by invertible matrices.

(ii) Show that the natural map

$$H^1\big(G, \mathrm{PGL}_n(L)\big) \to \mathrm{Br}(L/K)$$

arising from part (i) is injective and one has

$$\mathrm{Br}(L/K) = \bigcup_n H^1\big(G, \mathrm{PGL}_n(L)\big).$$

(iii) Construct a canonical embedding of groups

$$\lambda\colon\ \mathrm{Br}(L/K) \hookrightarrow H^2(G, L^*).$$

Hint. Show that the coboundary map associated with the exact sequence of G-modules

$$1 \to L^* \to \mathrm{GL}_n(L) \to \mathrm{PGL}_n(L) \to 1$$

defines the map of sets

$$\lambda\colon \operatorname{Br}(L/K) \to H^2(G, L^*)$$

and that $\lambda^{-1}(0) = \{0\}$. Note also that the tensor product of algebras corresponds to the map

$$H^1\big(G, \mathrm{PGL}_m(L)\big) \times H^1\big(G, \mathrm{PGL}_n(L)\big) \to H^1\big(G, \mathrm{PGL}_{mn}(L)\big)$$

induced by the tensor product of matrices

$$\mathrm{PGL}_m(L) \times \mathrm{PGL}_n(L) \to \mathrm{PGL}_{mn}(L)\,.$$

Furthermore, to prove that λ is a group homomorphism, use the explicit formula for the coboundary map from Exercise 1.1.7(iii). Finally, show that the homomorphism λ is injective.

(iv) Put $d = [L : K]$. Prove that the coboundary map

$$H^1\big(G, \mathrm{PGL}_d(L)\big) \to H^2(G, L^*)$$

is surjective.

Hint. Consider a vector space U over L with basis $\{\sigma_g \mid g \in G\}$. For an arbitrary 2-cocycle

$$\psi\colon G \times G \to L^*,$$

consider the map

$$\tilde{\varphi}\colon G \to \operatorname{Aut}(U) \cong \mathrm{GL}_d(L), \quad \tilde{\varphi}(g)\colon \sigma_h \mapsto \psi(g,h)\sigma_{gh}$$

and the corresponding 1-cocycle

$$\varphi\colon G \to \mathrm{PGL}_d(L);$$

cf. Exercise 1.1.7(iii). In a more invariant way, by Exercise 1.1.4(ii) the element $\alpha \in H^2(G, L^*)$ corresponds to a group extension

$$1 \to L^* \xrightarrow{\imath} \widetilde{G} \to G \to 1\,.$$

Consider the twisted group algebra $L * \widetilde{G}$: as an L-vector space it is isomorphic to the usual group algebra $L[\widetilde{G}]$, while multiplication in $L * \widetilde{G}$ is uniquely defined by the rule

$$g \cdot x = g(x)x \cdot g, \quad g \in \widetilde{G},\ x \in L\,,$$

where the action of $\widetilde{G}$ on L factors through the natural action of G on L. In particular, $L * \widetilde{G}$ is an algebra over K and L is not contained in its center. Now consider the quotient algebra

$$A = L * \widetilde{G}/\langle x \cdot e - 1 \cdot \imath(x)\rangle_{x \in L^*}\,,$$

where e denotes the neutral element in $\widetilde{G}$. Then A is a central simple algebra over K, and one has $\lambda([A]) = \alpha$.

(v) Show that

$$\lambda\colon \operatorname{Br}(L/K) \to H^2(G, L^*)$$

is an isomorphism.

EXERCISE 3.1.3. **Behavior with respect to field extensions** Consider a composition of finite Galois extensions

$$K \subset L \subset F$$

such that F is a Galois extension of the field K. Put

$$G = \mathrm{Gal}(L/K), \quad N = \mathrm{Gal}(F/L),$$

and $H = \mathrm{Gal}(F/K)$.

(i) Show that the following diagram commutes:

$$\begin{array}{ccc} \mathrm{Br}(L/K) & \longrightarrow & \mathrm{Br}(F/K) \\ \downarrow & & \downarrow \\ H^2(G, L^*) & \longrightarrow & H^2(H, F^*) \end{array}$$

Here the upper horizontal arrow is defined in a natural way, the vertical arrows are isomorphisms from Exercise 3.1.2(v), and the lower horizontal arrow is the inflation map, that is, the composition of the pull-back map

$$H^2(G, L^*) \to H^2(H, L^*)$$

defined by the group homomorphism $H \twoheadrightarrow G$ and the natural map

$$H^2(H, L^*) \to H^2(H, F^*).$$

Hint. Use Exercises 2.2.3(i) and 3.1.2(ii).

(ii) Construct a canonical isomorphism of groups

$$\mathrm{Br}(K) \cong H^2\big(G_K, (K^{sep})^*\big),$$

where, as before, G_K denotes the absolute Galois group of the field K.

Hint. Apply part (i).

(iii) Show that the following diagram commutes:

$$\begin{array}{ccc} \mathrm{Br}(F/K) & \longrightarrow & \mathrm{Br}(F/L) \\ \downarrow & & \downarrow \\ H^2(H, F^*) & \longrightarrow & H^2(N, F^*) \end{array}$$

Here the upper horizontal arrow is given by scalar extension from K to L for central simple algebras, the vertical arrows are isomorphisms from Exercise 3.1.2(v), and the lower horizontal arrow is the pull-back map defined by the homomorphism of groups $N \hookrightarrow H$.

Hint. Use Exercises 2.2.3(ii) and 3.1.2(ii).

Note that this commutative diagram can be constructed even if the field extension $K \subset L$ is not Galois.

(iv) Apply Exercise 3.1.2(v) to prove that there is an exact sequence

$$0 \to H^2(G, L^*) \to H^2(H, F^*) \to H^2(N, F^*).$$

Also deduce its existence from Exercise 1.2.6(iii) using Hilbert's Theorem 90.

In the notation of Exercise 3.1.3(iii), the embedding $N \hookrightarrow H$ also defines a direct image map

$$\mathrm{Br}(F/L) \cong H^2(N, F^*) \to H^2(H, F^*) \cong \mathrm{Br}(F/K);$$

see Exercise 1.2.7(ii). By Exercise 1.2.7(iii), the composition

$$\mathrm{Br}(F/K) \to \mathrm{Br}(F/L) \to \mathrm{Br}(F/K)$$

is the same as multiplication by $[L : K]$; we will use this in Exercise 3.1.10(i) below.

Note that the map $\mathrm{Br}(F/L) \to \mathrm{Br}(F/K)$ sends the class of a central simple algebra over L to the class of its Weil restriction from L to K, which is defined similarly to Weil restriction for commutative algebras, that is, for affine varieties over L; see §8 and Exercise 8.1.4 below. One can find more details in [**Tig87**].

EXERCISE 3.1.4. **Field $K = \mathbb{R}$ of real numbers**

(o) Produce a non-trivial element in the Brauer group $\mathrm{Br}(\mathbb{R})$.
(i) Show that

$$\mathrm{Br}(\mathbb{R}) \cong \mathbb{Z}/2\mathbb{Z}.$$

Hint. Use Exercise 1.1.5.

Compare this with the Frobenius theorem on the structure of finite-dimensional associative division algebras over $\mathbb{R}$.

EXERCISE 3.1.5. **Quaternion algebras**

Suppose that the characteristic of K is different from 2. Choose non-zero elements $a, b \in K$. Denote by $A(a, b)$ the quotient algebra

$$K\{i, j\}/\langle i^2 - a, j^2 - b, ij + ji\rangle$$

of the free associative algebra $K\{i, j\}$. This algebra is called the *quaternion algebra* corresponding to the elements a and b.

(i) Prove that $(1, i, j, ij)$ is a basis of $A(a, b)$ over K and $A(a, b)$ is a central simple algebra over K.
(ii) Show that $A(a, b) \cong \mathrm{M}_2(K)$ if and only if the equation

$$u^2 - av^2 - bw^2 = 0$$

has a non-zero solution in K.

Hint. First show that there is an isomorphism $A(a, b) \cong \mathrm{M}_2(K)$ if and only if $A(a, b)$ contains non-invertible elements. Then recall how one computes inverse elements in quaternion algebras. Finally, show that the equation

$$x^2 - ay^2 - bz^2 + abt^2 = 0$$

has a non-zero solution if and only if the equation

$$u^2 - av^2 - bw^2 = 0$$

has a non-zero solution, because both of these equations are equivalent to the requirement that b be contained in the image of the Galois norm for the field extension $K \subset K(\sqrt{a})$. Compare this argument with Exercise 6.1.2(ii) below.

(iii) Show that if a or b is a square in the field K, then $A(a,b) \cong \mathrm{M}_2(K)$. Also check that if $a+b=1$, then $A(a,b) \cong \mathrm{M}_2(K)$.

Hint. Use part (ii).

(iv) Prove that there is an isomorphism

$$A(a,b) \otimes_K A(a',b) \cong A(aa',b) \otimes_K \mathrm{M}_2(K)\,.$$

Hint. Choose the natural basis in the algebra $A(a,b) \otimes_K A(a',b)$ and consider its parts generating the subalgebras $A(a,b^2)$ and $A(aa',b)$.

Use this together with part (iii) to show that there is an isomorphism

$$A(a,b) \otimes_K A(a,b) \cong \mathrm{M}_4(K)\,.$$

The construction we present below generalizes Exercise 3.1.5.

EXERCISE 3.1.6. **Cyclic algebras**

Let $K \subset L$ be a finite Galois extension with a cyclic Galois group G. Fix a generator s of G.

(o) Show that the choice of s defines an isomorphism

$$\mathrm{Br}(L/K) \cong K^* / \mathrm{Nm}_{L/K}(L^*)\,,$$

where $\mathrm{Nm}_{L/K}$ denotes the Galois norm.

Hint. Apply Exercises 3.1.2(v) and 1.1.5(i).

(i) Show that under the isomorphism from part (o), the class of an element $a \in K^*$ in the group $K^*/\mathrm{Nm}_{L/K}(L^*)$ corresponds to the quotient algebra

$$A = L_G[\sigma]/\langle \sigma^n - a\rangle\,,$$

where $L_G[\sigma]$ is the L-vector space of polynomials in σ with multiplication given by the relation

$$\sigma \cdot x = {}^s x \cdot \sigma, \qquad x \in L\,.$$

Hint. Use Exercise 1.1.5(iv) and the hint for Exercise 3.1.2(iv).

Central simple algebras constructed in this way are called *cyclic*. Note that the algebra A depends not only on the cyclic field extension $K \subset L$ and the element $a \in K^*$, but also on the choice of the generator $s \in G$.

(ii) Suppose that the characteristic of K is coprime to n and K contains a primitive nth root of unity ζ. Kummer theory (see Exercise 2.4.2) tells us that $L = K(\sqrt[n]{b})$ for some element $b \in K^*$ and there is an isomorphism

$$G \cong \langle \zeta \rangle \subset K^*\,.$$

Choose $s = \zeta$ as generator of the group G. Show that there is an isomorphism of algebras

$$A \cong A_\zeta(a,b)\,,$$

where A is an algebra from part (i) and

$$A_\zeta(a,b) = K\{u,v\}/\langle u^n - a, v^n - b, uv - \zeta vu\rangle\,.$$

(iii) Under the assumptions of part (ii), show that the following equalities hold in the Brauer group $\mathrm{Br}(K)$:

$$[A_\zeta(a,b)] = -[A_\zeta(b,a)] = [A_{\zeta^{-1}}(b,a)]\,,$$
$$[A_\zeta(a,b)] + [A_\zeta(a',b)] = [A_\zeta(aa',b)]\,.$$

Moreover, one has $[A_\zeta(a,b)] = 0$ in $\mathrm{Br}(K)$ if and only if a is contained in the image of the Galois norm for the field extension $K \subset K(\sqrt[n]{b})$. Prove that

$$[A_\zeta(a, 1-a)] = 0$$

and that

$$[A_\zeta(a, -a)] = 0\,.$$

Hint. The first equality is immediately implied by part (ii). Bimultiplicativity and the assertion about the Galois norm follow from part (o). To prove the last two equalities, show that $1-a$ and $-a$ are norms with respect to the extension $K \subset K(\sqrt[n]{a})$.

DEFINITION 3.1.7. The *Milnor* K-*group* $\mathrm{K}_2(K)$ of a field K is defined as the quotient of the group $K^* \otimes_{\mathbb{Z}} K^*$ by the subgroup generated by all elements of the form $a \otimes (1-a)$, where $a \in K$, $a \neq 0, 1$. The image of $a \otimes b$ in the group $\mathrm{K}_2(K)$ is denoted by $\{a, b\}$. The relation

$$\{a, 1-a\} = 0$$

in the group $\mathrm{K}_2(K)$ is called the *Steinberg relation*. Homomorphisms from the group $\mathrm{K}_2(K)$ to an arbitrary (abelian) group are called *symbols*.

EXERCISE 3.1.8. **Relations in the group** K_2
Prove that in the group $\mathrm{K}_2(K)$ the following relations hold:

$$\{a, -a\} = 0, \quad \{a, b\} = -\{b, a\}\,.$$

Hint. To prove the first relation, compute the value of

$$a^{-1} \otimes (1 - a^{-1}) + a \otimes (1-a)$$

in the group $K^* \otimes_{\mathbb{Z}} K^*$. To prove the second relation, consider the element

$$\{ab, -ab\} \in \mathrm{K}_2(K)$$

and apply the first relation.

EXERCISE 3.1.9. **Norm residue symbol**
Suppose that the characteristic of K is coprime to n and K contains a primitive nth root of unity ζ. Show that the following is a well-defined group homomorphism:

$$\nu_\zeta\colon \mathrm{K}_2(K)/n \to \mathrm{Br}(K)_n, \quad \{a, b\} \mapsto [A_\zeta(a, b)]\,.$$

Hint. Use Exercise 3.1.6(iii).

This map is called the *norm residue symbol*.

The *Merkurjev–Suslin theorem* (see [**MS82**]) states that ν_ζ is an isomorphism. In particular, the fact that ν_ζ is surjective means that every element of order n in the Brauer group can be represented as a tensor product of cyclic algebras of dimension n^2.

EXERCISE 3.1.10. **Torsion in the Brauer group**

(o) Show that $\mathrm{Br}(K)$ is a torsion group.

Hint. Use Exercise 2.3.7(iv).

(i) Show that for a finite separable (but not necessarily normal) extension of fields $K \subset L$, all elements in the Brauer group $\mathrm{Br}(L/K)$ are annihilated by multiplication by the degree of the extension $[L : K]$.

Hint. Apply Exercise 1.2.7(iii) for the embedding of profinite groups

$$G_L \hookrightarrow G_K$$

and the G_K-module $(K^{sep})^*$, and also use Exercise 3.1.3(iii).

(ii) Suppose that an element $\alpha \in \mathrm{Br}(K)$ can be represented by a central simple algebra of dimension n^2 over K, where n is coprime to the characteristic of K. Show that the relation $n\alpha = 0$ holds in the group $\mathrm{Br}(K)$.

Hint. Use Exercise 3.1.2(i), the exact sequence of G_K-modules

$$1 \to \mu_n \to \mathrm{SL}_n(K^{sep}) \to \mathrm{PGL}_n(K^{sep}) \to 1\,,$$

where μ_n denotes the group of nth roots of unity in the field K^{sep}, and also Exercise 2.2.5(i).

(iii) Suppose that n is coprime to the characteristic of the field K. Show that there is an isomorphism

$$H^2(G_K, \mu_n) \xrightarrow{\sim} \mathrm{Br}(K)_n\,.$$

Hint. Use the Kummer exact sequence from Exercise 2.4.2(i).

Actually, one can show (see [**Ser79**, Exercise X.5.2]) that the assertion of Exercise 3.1.10(i) holds for an arbitrary (possibly non-separable) finite extension L of the field K. The assertion of Exercise 3.1.10(ii) also holds for an arbitrary n (possibly divisible by the characteristic of K). This follows from a generalization of Exercise 3.1.10(i) for the case of an arbitrary finite extension, from the fact that $\alpha \in \mathrm{Br}(L/K)$, where L is a maximal subfield in the division algebra D representing the element α, and from the equality $[L : K] = n$, where n^2 is the dimension of D over K.

Note that the converse assertion to Exercise 3.1.10(ii) does not hold: an arbitrary element of order n in the Brauer group is not necessarily representable by an algebra of dimension n^2. For instance, one can show (see [**GS06**, Example 1.5.7]) that for

$$K = \mathbb{C}(t_1, t_2, t_3, t_4),$$

the algebra

$$A = A(t_1, t_2) \otimes_K A(t_3, t_4)$$

defines an element of order 2 in $\mathrm{Br}(K)$ but is a division algebra of dimension 16 and thus is not representable by an algebra of dimension 4. Keeping in mind the Merkurjev–Suslin theorem and Exercise 3.3.2(ii), we can interpret this as the fact that the image of the element

$$\{t_1, t_2\} + \{t_3, t_4\}$$

in the quotient $\mathrm{K}_2(K)/2$ of $\mathrm{K}_2(K)$ is not an image of any decomposable element of $\mathrm{K}_2(K)$, that is, of an element of the form $\{a, b\}$.

DEFINITION 3.1.11. The *index* of an element $\alpha \in \mathrm{Br}(K)$ is the square root of the dimension of the central division algebra over K that represents α (note that the index is always an integer).

Exercise 3.1.10 together with the discussion before Definition 3.1.11 imply that the order of every element of the Brauer group divides its index. Note that the index of an element $\alpha \in \mathrm{Br}(K)$ equals the minimal degree of a separable extension L of K such that $\alpha \in \mathrm{Br}(L/K)$; see [**GS06**, Corollary 4.5.9]. Moreover, one can prove (see [**Ser79**, Exercise X.5.3b]) that the order and the index have the same set of prime divisors. A conjecture of Colliot-Thélène predicts that if K is a field of rational functions $k(X)$ on an irreducible variety X of dimension d over a separably closed field k, then the index of every element $\alpha \in \mathrm{Br}(K)$ divides the number m^{d-1}, where m is the order of α. This is equivalent to the following: every element of order m in the group $\mathrm{Br}(K)$ can be represented by a central simple algebra of dimension m^{2d-2}. For $d = 1$ this means that $\mathrm{Br}(K) = 0$. This particular case is well known, and later (in Exercise 3.2.3) we will give its proof. For $d = 2$ the conjecture of Colliot-Thélène means that every element of order m in the group $\mathrm{Br}(K)$ can be represented by a central division algebra of dimension m^2, or, equivalently, for every central division algebra of dimension n^2 over K, the order of its class in the Brauer group $\mathrm{Br}(K)$ equals n. This assertion was proved by de Jong in [**dJ04**], and two other proofs were given later in [**Lie08**] and [**SdJ10**]. Concerning the latter paper, see also the lecture notes [**Sta09**] and a generalization of its main result obtained in [**SX17**].

EXERCISE 3.1.12. **Reduced norm**
Let A be a central simple algebra of dimension n^2 over a field K.

(i) Show that the determinant map

$$\det\colon \mathrm{M}_n(K^{sep}) \to K^{sep}$$

is invariant under automorphisms of the algebra $\mathrm{M}_n(K^{sep})$.

Hint. Recall that all automorphisms of $\mathrm{M}_n(K^{sep})$ are conjugations by invertible matrices.

(ii) Consider an isomorphism

$$\chi\colon K^{sep} \otimes_K A \xrightarrow{\sim} \mathrm{M}_n(K^{sep})\,.$$

Prove that the composition

$$\det \circ \chi\colon K^{sep} \otimes_K A \to K^{sep}$$

commutes with the action of the Galois group G_K and does not depend on the choice of χ.

Hint. The difference between the actions of the Galois group on $K^{sep} \otimes_K A$ and on $\mathrm{M}_n(K^{sep})$, as well as between different choices of χ, amounts to an automorphism of the algebra $\mathrm{M}_n(K^{sep})$.

(iii) Prove that there exists a unique map of degree n from A to K (that is, a map given by a polynomial of degree n in coordinates of elements of A with respect to an arbitrary basis of A over K) such that after scalar extension from K to K^{sep} it corresponds to the determinant map

$$\det\colon \mathrm{M}_n(K^{sep}) \to K^{sep}\,.$$

This map is called the *reduced norm* and is denoted by

$$\mathrm{Nrd}\colon A \to K\,.$$

(iv) Show that for non-zero elements $a, b \in K$ and an element

$$\alpha = x + yi + zj + tij$$

of the quaternion algebra $A(a, b)$ one has

$$\mathrm{Nrd}(\alpha) = x^2 - ay^2 - bz^2 + abt^2\,.$$

(v) Show that A is a division algebra if and only if the hypersurface of degree n in $\mathbb{P}(A) \cong \mathbb{P}^{n^2-1}$ defined by the equation $\mathrm{Nrd}(\alpha) = 0$ has no points over K.

3.2. Brauer group and arithmetic properties of fields

DEFINITION 3.2.1. A field K *is of type* C_1 if every hypersurface $X_d \subset \mathbb{P}^n$ of degree d defined over K has a K-point provided that $d \leqslant n$.

Ignoring for simplicity the details concerning singularities, one can say that the C_1 condition means that every Fano hypersurface has a K-point.

EXERCISE 3.2.2. **Brauer group and C_1 condition**
Prove that $\mathrm{Br}(K) = 0$ for every field K of type C_1.
Hint. Apply Exercise 3.1.12(v).

EXERCISE 3.2.3. **Tsen's theorem**

(i) Let k be an algebraically closed field, and let t be a formal variable. Prove that the field $K = k(t)$ is of type C_1.

Hint. Let $f(x_0, \ldots, x_n)$ be a homogeneous polynomial of degree d with coefficients from K. We may assume that the coefficients are polynomials in t of degree at most c for some positive integer c. For every positive integer r we will look for a polynomial solution of the equation $f = 0$, of degree at most r in t, represented as

$$x_i = \sum_{j=0}^{r-1} y_{ij} t^j\,.$$

Show that in this setting the equation $f = 0$ over K is equivalent to a system of at most $c + dr$ homogeneous equations for $(n+1)r$ variables y_{ij} over k. Now it remains to consider a sufficiently large degree r.

(ii) Let K be a field of type C_1. Prove that every finite extension L of the field K is also of type C_1.

Hint. Choose a basis $(e_1, \ldots, e_r)$ in L over K. Given a homogeneous polynomial $f(x_0, \ldots, x_n)$ of degree d with coefficients from L, consider the polynomial

$$g(x_{01}, \ldots, x_{0r}, \ldots, x_{n1}, \ldots, x_{nr}) = \mathrm{Nm}_{L/K}\left(f(x_0, \ldots, x_n)\right)$$

of degree dr in $(n+1)r$ variables with coefficients from K, where

$$x_i = \sum_{j=1}^{r} x_{ij} e_j$$

and $\mathrm{Nm}_{L/K}\colon L \to K$ denotes the norm for the field extension $K \subset L$. Finally, observe that the equation $f = 0$ has a non-zero solution over L if and only if the equation $g = 0$ has a non-zero solution over K.

(iii) Prove *Tsen's theorem*: if k is an algebraically closed field, then every field K of transcendence degree 1 over k is of type C_1. In particular, one has $\mathrm{Br}(K) = 0$ (see Exercise 3.2.2).

EXERCISE 3.2.4. **Chevalley–Warning theorem**
Let $K = \mathbb{F}_q$ be a finite field with $q = p^s$ elements.

(i) Show that for every integer $0 \leqslant a < q - 1$ the equality

$$\sum_{x \in K} x^a = 0$$

holds in the field K.

Hint. For $a = 0$ use the fact that $q = 0$ in the field K, and keep in mind that $0^0 = 1$ by definition. For $a > 0$ choose $t \in K^*$ such that $t^a \neq 1$, and consider the sum $\sum\limits_{x \in K} (tx)^a$.

(ii) Prove that

$$\sum_{x \in K^{n+1}} P(x) = 0$$

for every polynomial P in $n+1$ variables over K provided that the degree of P is less than $(n+1)(q-1)$.

Hint. If P is a monomial, apply part (i).

(iii) Let $X \subset \mathbb{P}^n$ be a hypersurface of degree d defined over K, where $d \leqslant n$. Prove the *Chevalley–Warning theorem*:

$$|X(K)| \equiv 1 \pmod{p}.$$

Hint. Use the fact that

$$|X(K)|(q-1) + 1 \equiv \sum_{x \in K^{n+1}} P(x) \pmod{p},$$

where

$$P(x) = 1 - f(x)^{q-1}$$

and $f(x) = 0$ is the equation of the hypersurface X. Then apply part (ii).

In particular, a finite field is of type C_1.

(iv) Use part (iii) to show that the Brauer group of a finite field is trivial. Give another proof of this fact using the observation that the Galois group of every finite extension of a finite field is cyclic, together with Hilbert's Theorem 90 and Exercise 1.1.5(i).

One can generalize the Chevalley–Warning theorem. Specifically, *Esnault's theorem* (see [**Esn03**]) states that the number of points on every smooth Fano variety (or, more generally, on every smooth rationally connected variety) over a finite field $\mathbb{F}_q$ is equal to 1 modulo q. In particular, such a variety always has a point over $\mathbb{F}_q$.

3.3. Brauer group and Severi–Brauer varieties

The reader may have noticed that central simple algebras of dimension n^2 and Severi–Brauer varieties (see Definition 2.2.6) of dimension $n-1$ over a field K are canonically parameterized by the same set, namely $H^1\big(G_K, \mathrm{PGL}_n(K^{sep})\big)$; see Exercises 3.1.2(i) and 2.2.8(o). This interplay between algebras and varieties can be described as follows.

EXERCISE 3.3.1. **A Severi–Brauer variety that corresponds to a central simple algebra**

(i) Let V be an n-dimensional vector space over the field K. Show that the set of n-dimensional right ideals in $\mathrm{End}_K(V)$ is canonically identified with the projectivization $\mathbb{P}(V)$.

Hint. A line in V corresponds to the right ideal in $\mathrm{End}_K(V)$ that consists of all endomorphisms with image contained in this line.

(ii) Given a central simple algebra A over K of dimension n^2, consider the Grassmannian $\mathrm{G}(A,n)$ of linear subspaces of dimension n in A. Show that there is a closed subvariety $X \subset \mathrm{G}(A,n)$ defined over K such that X parameterizes right ideals $I \subset A$ of dimension n. Show that for every field extension $K \subset L$ the set of L-points $X(L)$ is canonically identified with the set of right ideals of dimension n in A_L.

Hint. Choose a basis $e_1, \ldots, e_{n^2}$ of A over K, and let

$$M_i \colon A \to A$$

be the operator of right multiplication by e_i. Then the n-dimensional subspace $I \subset A$ is a right ideal if and only if $M_i(I) \subset I$ for every $1 \leqslant i \leqslant n^2$. The fact that the latter condition defines a Zariski-closed subset can be easily checked in local coordinates on the Grassmannian.

(iii) Prove that X is the Severi–Brauer variety that corresponds to A in the way described above in terms of first cohomology.

Hint. Use part (i).

EXERCISE 3.3.2. **Central simple algebras of dimension** 4
Suppose that the characteristic of K is different from 2.

(i) Given a quaternion algebra $A(a,b)$, $a, b \in K^*$, show that the corresponding Severi–Brauer variety is isomorphic over K to a conic given by the equation

$$u^2 - av^2 - bw^2 = 0\,.$$

Hint. Consider the linear function on $A(a,b)$ given by the formula

$$x + yi + zj + tij \mapsto t\,.$$

Check that this function does not vanish identically on any two-dimensional right ideal I in $A(a,b)$ and that every two-dimensional right ideal I contains a unique (up to scaling) non-zero element of the form $x + yi + zj$. This element uniquely defines the ideal I because $A(a,b)$ does not contain one-dimensional right ideals. Conversely, given a point $(u : v : w)$ on the conic, consider a two-dimensional right ideal in $A(a,b)$

generated as a K-vector space by elements $u + vi + wj$ and $av + ui - wij$; cf. Exercise 6.1.2(ii).

(ii) Show that every central simple algebra of dimension 4 is a quaternion algebra.

Hint. Consider the corresponding Severi–Brauer variety.

The assertion of Exercise 3.3.2(ii) can also be proved in a direct algebraic way; see [**Bou58**, 11.2].

DEFINITION 3.3.3. Let X be a Severi–Brauer variety defined over the field K. The *class* of X in the Brauer group $\mathrm{Br}(K)$ is the class $b(X)$ of the corresponding central simple algebra.

EXERCISE 3.3.4. **Triviality of the class of a Severi–Brauer variety**

Let X be a Severi–Brauer variety defined over K. Prove that X has a K-point if and only if $b(X) = 0$ in the group $\mathrm{Br}(K)$.

Hint. Use Exercises 2.2.8(o),(i) and 3.1.2(ii).

Now we will give a more geometric description of the class $b(X)$. In order to do this, let us start with a description of classes in the Brauer group related to invariants in the Picard group. Recall that a variety X defined over a field K is said to be *geometrically irreducible* if the variety $X_{\bar{K}}$ is irreducible (and *geometrically reducible* otherwise).

Let X be a smooth projective geometrically irreducible variety over the field K. In particular, for every field extension $K \subset L$ there is a well-defined field of rational functions on X_L, which we will denote by $L(X)$. Note that if $K \subset L$ is a Galois extension, then there is an action of the Galois group $G = \mathrm{Gal}(L/K)$ on $L(X)$ so that for every function $f \in L(X)$ one has

$$g(f) = (g^{-1})^*(f),$$

where on the right-hand side we consider the action of G on the scheme X_L. In particular, the map

$$L(X)^* \to \mathrm{Div}(X_L)$$

that sends a rational function to its divisor commutes with the action of the group G.

EXERCISE 3.3.5. **Classes of invariants in the Picard group**

(i) Show that there is a canonical exact sequence

$$H^1\big(G_K, K^{sep}(X)^*/(K^{sep})^*\big) \to \mathrm{Br}(K) \to \mathrm{Br}\,\big(K(X)\big)$$

and a natural isomorphism

$$K(X)^*/K^* \cong \big(K^{sep}(X)^*/(K^{sep})^*\big)^{G_K}.$$

Hint. Consider the exact sequence of G_K-modules

$$1 \to (K^{sep})^* \to K^{sep}(X)^* \to K^{sep}(X)^*/(K^{sep})^* \to 1;$$

cf. Exercise 11.2.1(i). Use the fact that the canonical map

$$H^2\big(G_K, K^{sep}(X)^*\big) \to \mathrm{Br}\,\big(K(X)\big)$$

is injective, and also Exercise 2.1.3(v).

(ii) Show that there is a canonical exact sequence

$$0 \to \mathrm{Pic}(X) \to \mathrm{Pic}(X_{K^{sep}})^{G_K} \to H^1\big(G_K, K^{sep}(X)^*/(K^{sep})^*\big) \to 0\,.$$

Hint. Consider the exact sequence of G_K-modules

$$1 \to K^{sep}(X)^*/(K^{sep})^* \to \mathrm{Div}(X_{K^{sep}}) \to \mathrm{Pic}(X_{K^{sep}}) \to 0\,;$$

cf. Exercise 11.2.1(ii). Use part (i) together with Exercise 2.5.1(iii).

(iii) Using parts (i) and (ii), construct a canonical exact sequence

$$0 \to \mathrm{Pic}(X) \longrightarrow \mathrm{Pic}(X_{K^{sep}})^{G_K} \xrightarrow{\xi} \mathrm{Br}(K) \longrightarrow \mathrm{Br}\big(K(X)\big)\,.$$

(iv) Now we are going to give an explicit description of the map ξ from part (iii). Pick an element of the group $\mathrm{Pic}(X_{K^{sep}})^{G_K}$. It is defined over some finite Galois extension $K \subset L$ with Galois group G; that is, it is the class $[H]$ in the Picard group of a divisor H on X_L defined over L. Show that for every element $g \in G$ there is a rational function

$$f_g \in L(X)^*$$

on X_L such that its divisor equals $g(H) - H$. Use this to deduce that for all $g, h \in G$ the divisor of the rational function

$$\omega(g,h) = f_g \cdot g(f_h) \cdot f_{gh}^{-1}$$

is trivial, that is,

$$\omega(g,h) \in L^* \subset L(X)^*\,.$$

Prove that ω is a 2-cocycle. Moreover, show that its class in $H^2(G, L^*)$ does not depend on the choice of the divisor H and the function f_g, and is equal to $\xi([H])$.

Hint. Apply Exercise 1.1.6(i) and (ii).

EXERCISE 3.3.6. **The class of a Severi–Brauer variety**
Let X be a Severi–Brauer variety of dimension $n-1$.

(i) Let V be an n-dimensional vector space over the field K and let $\gamma \in \mathrm{PGL}(V)$ be an automorphism of the projective space $\mathbb{P}^{n-1} = \mathbb{P}(V)$. Let $\Pi \subset \mathbb{P}^{n-1}$ be a hyperplane over K defined by a linear function $l \in V^\vee$. Check that the divisor $\gamma(\Pi) - \Pi$ on $\mathbb{P}(V)$ is the divisor of the rational function

$$\tilde{\gamma}(l) \cdot l^{-1}\,,$$

where $\tilde{\gamma} \in \mathrm{GL}(V)$ is an arbitrary lifting of the element $\gamma \in \mathrm{PGL}(V)$ and the action of $\mathrm{GL}(V)$ on $V^\vee$ is given by the formula

$$\sigma(l) = (\sigma^\vee)^{-1}(l), \quad \sigma \in \mathrm{GL}(V)\,. \tag{3.1}$$

(ii) Show that the Galois group G_K acts trivially on the group

$$\mathrm{Pic}(X_{K^{sep}}) \cong \mathbb{Z}\,.$$

Therefore, the hyperplane

$$\Pi \subset \mathbb{P}^{n-1}_{K^{sep}}$$

corresponds to a divisor $H \subset X_{K^{sep}}$ that defines a class

$$[H] \in \mathrm{Pic}(X_{K^{sep}})^{G_K}\,.$$

Prove that

$$\xi([H]) = -b(X)\,,$$

where ξ is defined in Exercise 3.3.5(iii).

Hint. Let V be an n-dimensional vector space over K. Fix an isomorphism

$$\mathbb{P}(V)_L \cong X_L\,,$$

where $K \subset L$ is a finite Galois extension with Galois group G. Suppose that the hyperplane $\Pi \subset \mathbb{P}(V)_L$ is defined over K, that is, it is given by a linear function $l \in V^\vee$. Let

$$\varphi\colon G \to \mathrm{PGL}(V_L)$$

be a 1-cocycle and suppose that φ gives the form X of $\mathbb{P}(V)$ (see Exercise 2.2.1), defined by the isomorphism $\mathbb{P}(V)_L \cong X_L$ considered above. Then the divisor $g(H)$ corresponds to the hyperplane $\varphi(g)(\Pi)$. Now apply the explicit computation of the class $\xi([H])$ from Exercise 3.3.5(iv). Part (i) implies that the function f_g corresponds to the rational function on $\mathbb{P}^{n-1}_L$ given as the quotient of linear functions

$$\tilde{\varphi}(g)(l) \cdot l^{-1}\,,$$

where $\tilde{\varphi}(g) \in \mathrm{GL}(V)$ is an arbitrary lifting of the element $\varphi(g) \in \mathrm{PGL}(V)$. Therefore, one has

$$\omega(g,h) = \tilde{\varphi}(g)(l) \cdot l^{-1} \cdot {}^{g}\big(\tilde{\varphi}(h)\big)(l) \cdot \tilde{\varphi}(gh)(l)^{-1} \in L(X)^*\,,$$

where the rational function $\omega(g,h)$ is defined in Exercise 3.3.5(iv). It remains to apply formula (3.1) together with the observation that the operator

$$\tilde{\varphi}(g) \cdot {}^{g}\big(\tilde{\varphi}(h)\big) \cdot \tilde{\varphi}(gh)^{-1}\colon V \to V$$

is multiplication by a scalar, and use Exercise 1.1.7(iii).

(iii) Show that there is an exact sequence

$$0 \to \mathrm{Pic}(X) \to \mathbb{Z} \to \mathrm{Br}(K) \to \mathrm{Br}\big(K(X)\big)\,,$$

where the first map sends a divisor to its degree in

$$X_{K^{sep}} \cong \mathbb{P}^{n-1}_{K^{sep}}$$

and the second map sends 1 to $-b(X)$.

Hint. Use part (ii) together with Exercise 3.3.5(iii).

(iv) Prove that the order of the element $b(X) \in \mathrm{Br}(K)$ equals the minimal positive degree of a divisor on X that is defined over K.

Hint. Use part (iii).

(v) Show that the class $b(X)$ generates the kernel of the map

$$\mathrm{Br}(X) \to \mathrm{Br}\big(K(X)\big)\,.$$

In particular, if Severi–Brauer varieties X and Y are birational over K, then $b(X)$ and $b(Y)$ generate the same subgroup in $\mathrm{Br}(K)$.

Hint. Use part (iii).

Amitsur's conjecture (cf. [**Ami55**]) states that the converse assertion to Exercise 3.3.6(v) also holds. Specifically, if the classes $b(X)$ and $b(Y)$ of Severi–Brauer varieties of equal dimension over K generate the same subgroup in $\mathrm{Br}(K)$, then the varieties X and Y are birational over K.

Now we will give a more explicit solution for Exercise 3.3.6(v). Recall that every rational map between two projective spaces has a unique representation as a collection of homogeneous polynomials of equal degree without common divisors. The degree of these polynomials is called the degree of the corresponding rational map. Every linear rational map, that is, every rational map of degree one, is a composition of a linear embedding and a linear rational projection.

EXERCISE 3.3.7. **Morphisms between Severi–Brauer varieties**

(i) Show that for $m \leqslant n$ the forms of linear embeddings

$$\mathbb{P}^{m-1} \hookrightarrow \mathbb{P}^{n-1},$$

as well as the forms of linear rational projections

$$\mathbb{P}^{n-1} \dashrightarrow \mathbb{P}^{n-m-1},$$

are parameterized by the set $H^1\big(G_K, \mathrm{PGL}_{n,m}(K^{sep})\big)$, where

$$\mathrm{PGL}_{n,m}(K^{sep}) = \mathrm{GL}_{n,m}(K^{sep})/(K^{sep})^*,$$

with $\mathrm{GL}_{n,m}(K^{sep})$ consisting of all $n \times n$ matrices with entries from the field K^{sep} that have a zero lower left rectangle of size $(n-m) \times m$.

(ii) Let X and Y be Severi–Brauer varieties defined over K and let $\varphi \colon X \dashrightarrow Y$ be a rational map over K that becomes a linear rational projection between projective spaces after a scalar extension from K to K^{sep}. Show that

$$b(X) = b(Y).$$

Hint. If φ is a linear embedding or a linear rational projection, use part (i). Here one has to consider natural maps between exact sequences of G_K-modules,

$$\begin{array}{ccccccccc} 1 & \longrightarrow & (K^{sep})^* & \longrightarrow & \mathrm{GL}_{n,m}(K^{sep}) & \longrightarrow & \mathrm{PGL}_{n,m}(K^{sep}) & \longrightarrow & 1 \\ & & \big\downarrow{\scriptstyle \mathrm{id}} & & \big\downarrow & & \big\downarrow & & \\ 1 & \longrightarrow & (K^{sep})^* & \longrightarrow & \mathrm{GL}_r(K^{sep}) & \longrightarrow & \mathrm{PGL}_r(K^{sep}) & \longrightarrow & 1 \end{array}$$

for $r = n, m$, and $n-m$, which induce the identity map on $(K^{sep})^*$. Then apply Exercise 3.1.2(iii) to deduce the general case.

(iii) Let X and Y be Severi–Brauer varieties defined over K and let $\varphi \colon X \dashrightarrow Y$ be a rational map over K that becomes a rational map of degree d between projective spaces after a scalar extension from K to K^{sep}. Show that

$$d \cdot b(X) = b(Y).$$

Hint. First decompose an arbitrary rational map between projective spaces into a composition of a Veronese embedding and a linear rational map. Then observe that the dth symmetric power of the tautological representation of the group $\mathrm{GL}_n(K^{sep})$ defines a map between exact sequences of G_K-modules

$$\begin{array}{ccccccccc} 1 & \longrightarrow & (K^{sep})^* & \longrightarrow & \mathrm{GL}_n(K^{sep}) & \longrightarrow & \mathrm{PGL}_n(K^{sep}) & \longrightarrow & 1 \\ & & \downarrow{\scriptstyle d} & & \downarrow & & \downarrow & & \\ 1 & \longrightarrow & (K^{sep})^* & \longrightarrow & \mathrm{GL}_N(K^{sep}) & \longrightarrow & \mathrm{PGL}_N(K^{sep}) & \longrightarrow & 1 \end{array}$$

Here N stands for the binomial coefficient

$$N = \binom{n+d-1}{n},$$

and on $(K^{sep})^*$ the corresponding map is given by raising to the dth power. Finally, apply part (ii).

(iv) Solve Exercise 3.3.6(v) using the previous parts of the current exercise.

3.4. Further reading

A geometric treatment of Severi–Brauer varieties can be found in [**Kol16a**]. For partial results about function fields of Severi–Brauer varieties and on Amitsur's conjecture we refer the reader to [**Ami55**], [**Roq63**], [**Roq64**], [**Tre91**], [**Kra01**], and[**ABGV11**, §10]. Birational maps between Severi–Brauer surfaces were described in detail in [**Wei89**] and later (as a particular case of a much more general result) in [**IT91**] and [**Isk96**] (see also [**Cor05**]). More information on morphisms to Severi–Brauer varieties can be found in [**Lie17**]. For a relation between Severi–Brauer varieties and quadrics see [**Cla06**] and [**Lie17**]. Some results on birational properties of symmetric powers of Severi–Brauer varieties were obtained in [**KS04**] and [**Kol16b**].

CHAPTER 4

Residue Map on a Brauer Group

4.1. Complete discrete valuation fields

We start by recalling the main notions and facts concerning complete discrete valuation fields (for details see [**CF67**], [**Ser79**], and [**Lan64**]). A *discrete valuation* on a field K is a surjective group homomorphism

$$v\colon K^* \to \mathbb{Z}$$

such that the inequality

$$v(x+y) \geqslant \min\{v(x), v(y)\}$$

holds for all $x, y \in K$ (we put $v(0) = +\infty$ by definition). The *valuation ring* is the subring $\mathcal{O}_K \subset K$ given by the condition $v(x) \geqslant 0$. The *valuation ideal* is the ideal $\mathfrak{m}_K \subset K$ given by the condition $v(x) > 0$. The ring $\mathcal{O}_K$ is a local ring with maximal ideal $\mathfrak{m}_K$, which is generated by an arbitrary *uniformizer*, that is, an element $x \in K$ such that $v(x) = 1$. In particular, this implies the following important property of a discrete valuation: if $v(x) \neq v(y)$, then

$$v(x+y) = \min\{v(x), v(y)\}\,.$$

The *residue field* is the quotient of the valuation ring by its maximal ideal,

$$\kappa = \mathcal{O}_K/\mathfrak{m}_K\,.$$

If we fix a real number $\alpha > 1$, then a discrete valuation defines a metric on the field K by the formula

$$\rho(x,y) = \alpha^{-v(x-y)}, \quad x, y \in K\,.$$

A *complete discrete valuation field* is a discrete valuation field that is complete with respect to this metric. Completeness does not depend on the choice of α and is equivalent to the requirement that the natural map

$$\mathcal{O}_K \to \varprojlim \mathcal{O}_K/\mathfrak{m}_K^i$$

be an isomorphism. Examples of complete valuation fields include the field of p-adic numbers $\mathbb{Q}_p$ and the field of Laurent series $\kappa((u))$ (with obvious valuations in both cases).

In what follows, K denotes a complete discrete valuation field. One of the main arithmetic properties of the field K is *Hensel's lemma* (see [**Lan64**, § II.2]): let $f \in \mathcal{O}_K[t]$ be a polynomial whose reduction $\bar{f}$ modulo $\mathfrak{m}_K$ has a root $\bar{x}$ in the residue field κ such that $\bar{f}'(\bar{x}) \neq 0$; then there is a unique root $x \in \mathcal{O}_K$ of the polynomial f such that

$$x \equiv \bar{x} \pmod{\mathfrak{m}_K}\,.$$

If the characteristics of the fields K and κ are equal to each other, then the natural ring homomorphism $\mathcal{O}_K \to \kappa$ has a (non-canonical) section.

Suppose that the residue field κ has characteristic $p > 0$ and is perfect (that is, all irreducible polynomials over κ are separable). Then the natural group homomorphism

$$\mathcal{O}_K^* \to \kappa^*$$

has a canonical section $x \mapsto [x]$; see [**Ser79**, Proposition II.8]. The element $[x]$ is called the *Teichmuller representative* of the element x.

Choose a finite extension L of the field K. There is a unique discrete valuation

$$w\colon L^* \to \mathbb{Z}$$

such that $w|_K$ is proportional to the valuation v, that is, one has

$$w|_K = e \cdot v$$

for some positive integer e (see [**CF67**, II.10]); the number e is called the *ramification index* of the extension $K \subset L$. Moreover, the field L is complete with respect to the discrete valuation w, and the ring $\mathcal{O}_L$ is the integral closure of the ring $\mathcal{O}_K$ in L; see [**Ser79**, Proposition II.3]. One has $n = ef$, where $n = [L : K]$ and $f = [\lambda : \kappa]$, with $\lambda = \mathcal{O}_L/\mathfrak{m}_L$ being the residue field of L; see [**CF67**, Proposition I.5.3].

The extension $K \subset L$ is *unramified* if $e = 1$ (or, equivalently, if $n = f$) and the extension $\kappa \subset \lambda$ is separable. In this case we will denote a discrete valuation on the field L by the same symbol as the corresponding discrete valuation on the field K. Since $e = 1$ this does not lead to a contradiction. Unramified extensions are separable; see [**CF67**, Proposition I.7.1]. The category of unramified extensions of the field K is equivalent to the category of separable extensions of its residue field κ; see [**CF67**, Theorem I.7.1]. An example of an unramified extension is an extension $\mathbb{Q}_p \subset \mathbb{Q}_p(\sqrt{a})$ where $(a, p) = 1$ and $p \neq 2$, or an extension $\kappa((u)) \subset \lambda((u))$ where $\kappa \subset \lambda$ is a finite separable extension of fields.

The extension $K \subset L$ is *totally ramified* if $f = 1$, that is, if $\kappa = \lambda$ or, equivalently, if $n = e$. The extension $K \subset L$ is totally ramified if and only if $L = K(y)$, where y is a root of an *Eisenstein polynomial* $f(t) \in \mathcal{O}_K[t]$, that is, a polynomial

$$f(t) = t^n + a_{n-1}t^{n-1} + \ldots + a_1 t + a_0$$

such that $a_i \in \mathfrak{m}_K$ for all $i = 0, \ldots, n-1$ and $v(a_0) = 1$; see [**CF67**, Theorem I.6.1]. An example of a totally ramified extension is an extension $\mathbb{Q}_p \subset \mathbb{Q}_p(y)$ where y is a root of the polynomial $y^2 + p^2 y + p$, or an extension $\kappa((u)) \subset \kappa((v))$ where $v^n = u$.

For an arbitrary finite extension $K \subset L$ there exists the largest subfield $E \subset L$ that contains K and is unramified over it; see [**CF67**, Theorem I.7.2]. The residue field of E coincides with the separable closure of the field κ in the residue field λ of L. If the extension $\kappa \subset \lambda$ is separable, then the extension $E \subset L$ is totally ramified, and one has

$$[E : K] = f\,, \quad [L : E] = e\,.$$

Now let $K \subset L$ be a finite Galois extension with Galois group

$$G = \mathrm{Gal}(L/K)\,,$$

and suppose that the extension of the residue field $\kappa \subset \lambda$ is separable. Then the extension $\kappa \subset \lambda$ is also Galois; see [**CF67**, p. 27]. Put

$$G^{nr} = \mathrm{Gal}(\lambda/\kappa)\,.$$

We have a canonical homomorphism

$$\psi\colon G \to G^{nr}$$

defined as follows. Every automorphism $g \in G$ of the field L over K preserves the discrete valuation w on L compatible with the discrete valuation v on K, because w is unique (see above). Hence g preserves the subring $\mathcal{O}_L$ of L and also the maximal ideal $\mathfrak{m}_L \subset \mathcal{O}_L$. Taking g modulo the ideal $\mathfrak{m}_L$, we obtain an automorphism $\psi(g)$ of the field λ over κ. The homomorphism ψ is always surjective. The extension $K \subset L$ is unramified if and only if ψ is an isomorphism. The largest subfield $E \subset L$ unramified over K is the Galois extension of K that corresponds via Galois theory to the subgroup $\mathrm{Ker}(\psi)$. There is a canonical isomorphism

$$\mathrm{Gal}(E/K) \cong G^{nr}.$$

Let $K^{nr} \subset K^{sep}$ denote the largest subextension in K^{sep} unramified over K. There is a canonical isomorphism

$$\mathrm{Gal}(K^{nr}/K) \cong G_\kappa,$$

where as before G_κ is the absolute Galois group of the field κ.

EXERCISE 4.1.1. **Totally ramified extensions are cyclic**

Assuming the above notation, suppose that the field κ (and thus also the field K) has characteristic zero and that κ contains all roots of unity. Denote the group of all roots of unity by μ.

(i) Show that the field K also contains all roots of unity.

Hint. Apply Hensel's lemma.

(ii) Suppose that the extension $K \subset L$ is totally ramified and $L = K(y)$, where y is a root of an Eisenstein polynomial

$$f(t) = t^n + a_{n-1}t^{n-1} + \ldots + a_1 t + a_0 \in \mathcal{O}_K[t].$$

Prove that

$$L = K\left(\sqrt[n]{-a_0}\right).$$

Hint. Show that the element

$$z = \frac{a_{n-1}y^{n-1} + \ldots + a_1 y + a_0}{a_0} \in L$$

is contained in $\mathcal{O}_L$ and satisfies the condition

$$z \equiv 1 \pmod{\mathfrak{m}_L}.$$

Then use Hensel's lemma to show that z is an nth power in the field L. Note that the polynomial $t^n - z$ satisfies the assumptions of Hensel's lemma, because κ is of characteristic zero.

(iii) Using Kummer theory (see Exercise 2.4.2), show that every totally ramified finite extension $K \subset L$ is cyclic and that its Galois group is canonically embedded into μ.

(iv) Prove that for an arbitrary finite Galois extension $K \subset L$ the subgroup

$$\mathrm{Gal}(L/E) \subset G$$

is cyclic and central, where E is the largest subfield in L unramified over K.

Hint. Use the fact that μ is a trivial G-module.

In what follows, we will also need the following technical result. Let $K \subset L$ a finite separable extension of fields, v a discrete valuation on the field K, and K_v the completion of K with respect to v (in particular, unlike the above notation, the field K is not a *complete* discrete valuation field). Let W denote the set of discrete valuations w on the field L extending the valuation v, that is, such that the function $w|_K$ is proportional to v. For every $w \in W$, let L_w denote the completion of the field L with respect to the valuation w. Put $A = L \otimes_K K_v$.

EXERCISE 4.1.2. **Extension of a discrete valuation**

(i) Check that there is an isomorphism of K_v-algebras

$$A \cong \prod_{i \in I} L_i\,,$$

where I is a finite set and L_i, $i \in I$, are finite extensions of the field K_v.

Hint. By the primitive element theorem the field L is generated over K by one element. Decompose the minimal polynomial of this element into irreducible factors over the field K_v.

(ii) Show that for every $w \in W$ there is a uniquely defined homomorphism of K_v-algebras $A \to L_w$.

Hint. The required homomorphism is defined by the natural embedding $L \subset L_w$ and by the induced embedding $K_v \subset L_w$.

This gives a map $W \to I$.

(iii) Construct the inverse map $I \to W$.

Hint. For every finite extension of fields $K_v \subset L_i$ there is a unique extension of the valuation v from K_v to L_i. Define w as its restriction to $L \subset L_i$.

Therefore, there is a bijection $I \cong W$ and an isomorphism of K_v-algebras

$$A \cong \prod_{w \in W} L_w\,.$$

(iv) Suppose that $K \subset L$ is a Galois extension with Galois group G. The group G acts naturally on the set W. Prove that this action is transitive.

Hint. Otherwise one would have $A \cong B \times C$, where B and C are K_v-algebras with an action of the group G, which contradicts the equality $A^G = K_v$.

(v) Prove that for every $w \in W$, the extension $K_v \subset L_w$ is Galois and the natural homomorphism $\mathrm{Gal}(L_w/K_v) \to G$ induces an isomorphism $\mathrm{Gal}(L_w/K_v) \cong H$, where $H = \mathrm{Stab}_G(w)$.

Hint. There is a natural homomorphism $H \to \mathrm{Gal}(L_w/K_v)$. Check that it is injective and then use part (iv) to compare $|H|$ with $[L_w : K_v]$.

4.2. Brauer group of a complete discrete valuation field

Let K be a complete discrete valuation field with valuation ring $\mathcal{O}_K$. Suppose that the residue field κ is perfect. In this case *Lang's theorem* (see [**Lan52**] or [**Ser79**, § X.7]) states that the field K^{nr} is of type C_1 (note that since κ is perfect, the residue field of the field K^{nr} is algebraically closed). In particular, one has

$$\mathrm{Br}(K^{nr}) = 0$$

by Exercise 3.2.2. The latter can also be proved by applying certain finite group cohomology techniques together with the fact that for any composition of separable extensions

$$K^{nr} \subset E \subset F,$$

the norm map

$$\mathrm{Nm}_{F/E} \colon F^* \to E^*$$

is surjective (this is a non-trivial arithmetic result; see [**Ser79**, Proposition X.11] and [**Ser79**, Proposition V.7]).

EXERCISE 4.2.1. **Residue map on the Brauer group**

(i) Show that the natural map

$$\mathrm{Br}(K^{nr}/K) \to \mathrm{Br}(K)$$

is an isomorphism.

Hint. Recall that $\mathrm{Br}(K^{nr}) = 0$.

(ii) Construct a natural map

$$\mathrm{res} \colon \mathrm{Br}(K) \to \mathrm{Hom}(G_\kappa, \mathbb{Q}/\mathbb{Z}).$$

This is called the *residue map*.

Hint. Use the isomorphism

$$\mathrm{Br}(K^{nr}/K) \cong \mathrm{Br}(K)$$

from part (i), the map

$$H^2\big(G_\kappa, (K^{nr})^*\big) \to H^2(G_\kappa, \mathbb{Z})$$

that arises from the map of G_κ-modules $v \colon (K^{nr})^* \to \mathbb{Z}$, and also Exercise 1.3.3(ii).

(iii) Let $K \subset L$ be an unramified finite Galois extension with Galois group G, and let $\alpha \in H^2(G, \mathcal{O}_L^*)$. There are natural maps

$$H^2(G, \mathcal{O}_L^*) \to H^2(G, L^*) \xrightarrow{\sim} \mathrm{Br}(L/K) \hookrightarrow \mathrm{Br}(K),$$

as shown in Exercise 3.1.2(v). Therefore, there is a well-defined residue $\mathrm{res}(\alpha)$. Show that $\mathrm{res}(\alpha) = 0$.

In other words, the assertion of Exercise 4.2.1(i) means that for any central division algebra D of dimension n^2 over K, there is an unramified extension $K \subset E$ such that

$$E \otimes_K D \cong \mathrm{M}_n(E).$$

This can also be proved in a straightforward algebraic way, by showing that among maximal subfields of D there is a field E unramified over K (see the "Appendix" in [**CF67**, VI.1]).

Note that an element of $\mathrm{Hom}(G_\kappa, \mathbb{Q}/\mathbb{Z})$ corresponds to a cyclic extension of the field κ with a chosen generator of the Galois group.

DEFINITION 4.2.2. For non-zero elements $a, b \in K$, the *Hilbert symbol* is defined by the formula

$$(a, b)_v = (-1)^{v(a)v(b)}\, a^{-v(b)} b^{v(a)} \pmod{\mathfrak{m}_K} \in \kappa^*.$$

Exercise 4.2.3. **Residue for cyclic algebras**

(i) Show that the Hilbert symbol defines a group homomorphism

$$(-,-)_v \colon \mathrm{K}_2(K) \to \kappa^*, \quad \{a,b\} \mapsto (a,b)_v\,;$$

see Definition 3.1.7 for $\mathrm{K}_2(K)$.

(ii) Let $K \subset L$ be an unramified finite cyclic Galois extension of degree n with Galois group G, and let $a \in K^*$. Choose a generator s of the group G. Let A be a cyclic algebra over K that corresponds to these data (see Exercise 3.1.6(i)). Show that if $v(a) = 0$, then $\mathrm{res}([A]) = 0$.

Hint. Use Exercise 4.2.1(iii).

(iii) Suppose that in the notation of part (ii) one has $v(a) = 1$. Prove that $\mathrm{res}([A])$ corresponds to a cyclic extension $\kappa \subset \lambda$ with the generator

$$s \in G \cong \mathrm{Gal}(\lambda/\kappa)$$

chosen above.

Hint. Use functoriality of isomorphisms from Exercise 1.1.5(i) with respect to the morphism of G-modules $v\colon L^* \to \mathbb{Z}$. This gives a commutative diagram

$$\begin{array}{ccc} K^*/\,\mathrm{Nm}_{L/K}(L^*) & \longrightarrow & H^2(G,L^*) \\ \big\downarrow{\scriptstyle v} & & \big\downarrow \\ \mathbb{Z}/n\mathbb{Z} & \longrightarrow & H^2(G,\mathbb{Z}) \end{array}$$

Then apply Exercise 1.3.3(iii).

(iv) Choose an nth root of unity $\zeta \in \kappa$, where n is coprime to the characteristic of the field κ (and thus also to the characteristic of the field K). By Kummer theory (see Exercise 2.4.2) we get an isomorphism

$$\mathrm{Hom}(G_\kappa, \mathbb{Z}/n\mathbb{Z}) \cong \kappa^*/(\kappa^*)^n\,.$$

For simplicity we will also denote by res the composition

$$\mathrm{Br}(K)_n \xrightarrow{\mathrm{res}} \mathrm{Hom}(G_\kappa, \mathbb{Z}/n\mathbb{Z}) \cong \kappa^*/(\kappa^*)^n\,.$$

Prove that the following diagram is commutative (see Exercise 3.1.9):

$$\begin{array}{ccc} \mathrm{K}_2(K) & \xrightarrow{\nu_\zeta} & \mathrm{Br}(K)_n \\ \big\downarrow{\scriptstyle (-,-)_v} & & \big\downarrow{\scriptstyle \mathrm{res}} \\ \kappa^* & \longrightarrow & \kappa^*/(\kappa^*)^n \end{array}$$

Therefore, the residue of the class of the cyclic algebra $A_\zeta(a,b)$ corresponds to the Kummer extension

$$\kappa \subset \kappa\big(\sqrt[n]{(a,b)_v}\big)\,.$$

Hint. Recall that there is a (non-canonical) isomorphism

$$K^* \cong \mathbb{Z} \times \mathcal{O}_K^*\,.$$

Using Exercise 3.1.8, deduce from this fact that the group $\mathrm{K}_2(K)$ is generated by symbols $\{a,b\}$, where $v(b) = 0$ and $v(a)$ equals either 0 or 1. Then apply parts (ii) and (iii).

Our definition of the map res implies that for an unramified extension $K \subset L$ and an element $\alpha \in \mathrm{Br}(L/K)$, the residue $\mathrm{res}(\alpha)$ is the image of α with respect to the composition

$$H^2(G, L^*) \to H^2(G, \mathbb{Z}) \xrightarrow{\sim} \mathrm{Hom}(G, \mathbb{Q}/\mathbb{Z}) \cong \mathrm{Hom}(G^{nr}, \mathbb{Q}/\mathbb{Z}) \hookrightarrow \mathrm{Hom}(G_\kappa, \mathbb{Q}/\mathbb{Z}).$$

We will also use the notation res for the corresponding map

$$\mathrm{Br}(L/K) \to \mathrm{Hom}(G^{nr}, \mathbb{Q}/\mathbb{Z}).$$

EXERCISE 4.2.4. **Brauer group of a complete discrete valuation field** In parts (i) and (ii) we assume that the extension $K \subset L$ is an unramified finite Galois extension and $G = \mathrm{Gal}(L/K)$.

(i) Show that there is an isomorphism

$$\mathrm{Ker}\left(\mathrm{Br}(L/K) \xrightarrow{\mathrm{res}} \mathrm{Hom}(G^{nr}, \mathbb{Q}/\mathbb{Z})\right) \cong H^2(G, \mathcal{O}_L^*)$$

and that a choice of a uniformizer $\pi \in \mathcal{O}_K$ defines a splitting

$$\mathrm{Br}(L/K) \cong H^2(G, \mathcal{O}_L^*) \oplus \mathrm{Hom}(G^{nr}, \mathbb{Q}/\mathbb{Z}).$$

Hint. Use the fact that π defines a splitting of the exact sequence of G-modules

$$1 \to \mathcal{O}_L^* \to L^* \xrightarrow{v} \mathbb{Z} \to 0$$

and that $G \cong G^{nr}$.

(ii) The morphism of G-modules

$$\mathcal{O}_L^* \to \lambda^*, \quad x \mapsto x \pmod{\mathfrak{m}_L}$$

defines a map

$$H^2(G, \mathcal{O}_L^*) \to H^2(G, \lambda^*).$$

Show that it is an isomorphism.

Hint. Consider a decreasing filtration on $\mathcal{O}_L^*$ given by the subgroups $1+\mathfrak{m}_K^i$, $i \geqslant 1$. Since the adjoint quotients of this filtration are isomorphic to λ, by Exercise 2.2.4(iii) the filtration satisfies all the assumptions of Exercise 2.7.2(ii).

(iii) Show that there is an exact sequence

$$0 \to \mathrm{Br}(\kappa) \to \mathrm{Br}(K) \to \mathrm{Hom}(G_\kappa, \mathbb{Q}/\mathbb{Z}) \to 0$$

and that a choice of a uniformizer $\pi \in \mathcal{O}_K$ gives its splitting.

Hint. Use parts (i) and (ii), take a limit over all unramified extensions of the field K, and apply Exercise 4.2.1(i) and (ii).

(iv) Show that for an arbitrary separable finite extension $K \subset L$ (not necessarily normal or unramified) the following diagram commutes:

$$\begin{array}{ccc} \mathrm{Br}(K) & \xrightarrow{\mathrm{res}} & \mathrm{Hom}(G_\kappa, \mathbb{Q}/\mathbb{Z}) \\ \downarrow & & \downarrow e \\ \mathrm{Br}(L) & \xrightarrow{\mathrm{res}} & \mathrm{Hom}(G_\lambda, \mathbb{Q}/\mathbb{Z}) \end{array}$$

Here the right vertical arrow is a composition of the pull-back map

$$\mathrm{Hom}(G_\kappa, \mathbb{Q}/\mathbb{Z}) \to \mathrm{Hom}(G_\lambda, \mathbb{Q}/\mathbb{Z})$$

and multiplication by the ramification index e.

DEFINITION 4.2.5. A *non-archimedean local field* is a complete discrete valuation field with finite residue field (cf. the definition of a local field from Chapter 10).

One can show that every non-archimedean local field is either a finite extension of the field $\mathbb{Q}_p$ for some prime number p, or a field $\mathbb{F}_q((T))$ where q is a power of a prime.

EXERCISE 4.2.6. **Brauer group of a non-archimedean local field**
Let K be a non-archimedean local field.

(i) Prove that there is a canonical isomorphism

$$\mathrm{res}\colon \mathrm{Br}(K) \xrightarrow{\sim} \mathbb{Q}/\mathbb{Z}.$$

Hint. Use Exercise 4.2.4(iii), the Chevalley–Warning theorem from Exercise 3.2.4, and the fact that the Galois group of a finite field is canonically isomorphic to $\widehat{\mathbb{Z}}$.

(ii) Let $K \subset L$ be a finite extension of non-archimedean local fields. Show that the natural map

$$\mathrm{Br}(K) \to \mathrm{Br}(L)$$

corresponds to multiplication by the degree of the field extension

$$\mathbb{Q}/\mathbb{Z} \xrightarrow{n} \mathbb{Q}/\mathbb{Z}, \quad n = [L : K].$$

Hint. Apply Exercise 4.2.4(iv).

Let $\mu(K) \subset K^*$ be the group of roots of unity contained in the field K (we do not assume that K contains *all* roots of unity), and let $K \subset L$ be (as before) a finite Galois extension with Galois group G. For simplicity we will use the notation res also for the composition of the natural map

$$H^2\big(G, \mu(K)\big) \to H^2(G, L^*)$$

that arises from the embedding of G-modules $\mu(K) \hookrightarrow L^*$ and the usual residue map

$$\mathrm{res}\colon H^2(G, L^*) = \mathrm{Br}(L/K) \to \mathrm{Hom}(G_\kappa, \mathbb{Q}/\mathbb{Z}).$$

EXERCISE 4.2.7. **Kernel of the residue map**
Suppose that an element $\alpha \in H^2\big(G, \mu(K)\big)$ is contained in the image of the natural map

$$H^2\big(G^{nr}, \mu(K)\big) \to H^2\big(G, \mu(K)\big).$$

Prove that $\mathrm{res}(\alpha) = 0$.

Hint. Use the fact that the map res factors through the isomorphism

$$H^2(G^{nr}, E^*) \xrightarrow{\sim} H^2(G, L^*)$$

and that $v\big(\mu(K)\big) = 0$.

4.3. Unramified Brauer group of a function field

Now let K be a finitely generated field over a field k of characteristic zero. For every discrete valuation v on K, let K_v denote the completion of K with respect to v, and let κ_v denote the residue field of K_v. Suppose that the valuation v is trivial on the field k, that is, $v(k^*) = 0$. Then the field κ_v has characteristic zero as well; in particular, it is perfect. Denote by res_v the composition of the natural map

$$\mathrm{Br}(K) \to \mathrm{Br}(K_v)$$

with the residue map (see Exercise 4.2.1(ii))

$$\mathrm{Br}(K_v) \to \mathrm{Hom}(G_{\kappa_v}, \mathbb{Q}/\mathbb{Z})\,.$$

Since the field κ_v is perfect, the residue res_v is well-defined.

Note that if $\mathrm{char}(k) = p > 0$, then the residue is still well-defined (and all assertions in this section and in §4.4 below hold as well) after we replace the Brauer group by its subgroup that consists of all elements of order coprime to p. In this case, one defines the residue with the help of étale cohomology; see [**CT95**, §3.3] for more details.

If $K \cong k(X)$, where X is a normal irreducible variety over the field k, and D is a prime divisor on X, then we have a discrete valuation

$$v_D(f) = \mathrm{ord}_D(f), \quad f \in K^*\,,$$

such that $\kappa_{v_D} \cong k(D)$. Discrete valuations on the field K that are obtained in this way for some choice of the variety X and the divisor D are called *divisorial* valuations. For such valuations we abbreviate res_{v_D} to res_D.

Note that there exist non-divisorial discrete valuations. A discrete valuation v is divisorial if and only if its valuation ring in K is a localization of a finitely generated k-algebra (see e.g. [**Gab98**, Proposition 1]). The latter condition is also equivalent to

$$\mathrm{tr.\,deg}(K/k) - 1 = \mathrm{tr.\,deg}(\kappa_v/k)\,,$$

where tr. deg denotes the transcendence degree of a field extension.

DEFINITION 4.3.1. The *unramified Brauer group* of the field K is a subgroup

$$\mathrm{Br}^{nr}(K) = \bigcap_v \mathrm{Ker}(\mathrm{res}_v)$$

in the group $\mathrm{Br}(K)$, where v varies over the set of all discrete valuations on the field K that are trivial on k.

One can show that in the definition of the unramified Brauer group it is enough to let v vary over the set of divisorial valuations; see [**CT95**, Proposition 2.1.8e] and [**CT95**, §2.2.2].

The following exercise shows that the unramified Brauer group is constant under purely transcendental extensions.

EXERCISE 4.3.2. **Faddeev's theorem on the unramified Brauer group of a purely transcendental extension**

Let t be a variable, $\bar{K}$ the algebraic closure of the field K, and $\mathrm{Div}(\mathbb{A}^1_{\bar{K}})$ the group of divisors on the affine line $\mathbb{A}^1_{\bar{K}}$ over $\bar{K}$.

(i) Show that the natural map

$$\mathrm{Br}\left(\bar{K}(t)/K(t)\right) \to \mathrm{Br}\left(K(t)\right)$$

is an isomorphism.

Hint. Apply Tsen's theorem.

(ii) Prove that

$$H^1\left(G_K, \mathrm{Div}(\mathbb{A}^1_{\bar{K}})\right) = 0, \quad H^2\left(G_K, \mathrm{Div}(\mathbb{A}^1_{\bar{K}})\right) \cong \bigoplus_{x\in\mathbb{A}^1_K} \mathrm{Hom}\left(G_{K(x)}, \mathbb{Q}/\mathbb{Z}\right),$$

where the direct sum is taken over all closed schematic points $x \in \mathbb{A}^1_K$ and $K(x)$ denotes the residue field at the point x.

Hint. Apply Exercise 2.5.1(iii).

(iii) Consider the exact sequence of G_K-modules

$$1 \to \bar{K}^* \to \bar{K}(t)^* \to \mathrm{Div}(\mathbb{A}^1_{\bar{K}}) \to 0.$$

Use it to construct the exact sequence

$$0 \to \mathrm{Br}(K) \xrightarrow{\theta} \mathrm{Br}\left(K(t)\right) \xrightarrow{\oplus_x \mathrm{res}_x} \bigoplus_{x\in\mathbb{A}^1_K} \mathrm{Hom}\left(G_{K(x)}, \mathbb{Q}/\mathbb{Z}\right),$$

where θ is the natural map that arises from the extension of fields $K \subset K(t)$.

Hint. Use the isomorphism

$$G_K \cong \mathrm{Gal}\left(\bar{K}(t)/K(t)\right)$$

and also parts (i) and (ii).

(iv) Show that the image of the subgroup

$$\mathrm{Br}^{nr}(K) \subset \mathrm{Br}(K)$$

under the embedding θ is contained in the subgroup

$$\mathrm{Br}^{nr}\left(K(t)\right) \subset \mathrm{Br}\left(K(t)\right).$$

Hint. One can restrict a discrete valuation

$$w\colon K(t)^* \to \mathbb{Z}$$

to the subfield $K \subset K(t)$.

(v) Prove that every discrete valuation

$$v\colon K^* \to \mathbb{Z}$$

is a restriction of some discrete valuation

$$w\colon K(t)^* \to \mathbb{Z}.$$

Hint. Consider the valuation ring $\mathcal{O} \subset K$ and a uniformizer $\varpi \in \mathcal{O}$ associated with v. Prove that the localization of the ring $\mathcal{O}[t]$ at the prime ideal generated by ϖ is a valuation ring for some discrete valuation

$$w\colon K(t)^* \to \mathbb{Z}$$

that satisfies the required condition $w|_K = v$.

(vi) Prove that there is an embedding

$$\theta\big(\operatorname{Br}(K)\big) \cap \operatorname{Br}^{nr}\big(K(t)\big) \subset \theta\big(\operatorname{Br}^{nr}(K)\big).$$

Hint. Use part (v).

(vii) Using parts (iii), (iv), and (vi), prove *Faddeev's theorem*: the map θ induces an isomorphism

$$\operatorname{Br}^{nr}(K) \xrightarrow{\sim} \operatorname{Br}^{nr}\big(K(t)\big).$$

The assumption that the field k (and thus also the field K) has characteristic zero is used to apply Tsen's theorem in Exercise 4.3.2(i). To drop this assumption one has to assume that the field K is perfect.

One can find in [**Ste84**] an explicit description of the Brauer group $\operatorname{Br}\big(\mathbb{C}(t_1, \ldots, t_n)\big)$ as a direct sum of continuously many copies of $\mathbb{Q}/\mathbb{Z}$. The argument is very much in the spirit of the solution of Exercise 4.3.2.

DEFINITION 4.3.3. Irreducible varieties X and Y over a field k are *stably birationally equivalent*, or just *stably birational*, if for some m and n the varieties $X \times \mathbb{P}^m$ and $Y \times \mathbb{P}^n$ are birational over k. A variety which is stably birational to a point is said to be *stably rational*.

Exercise 4.3.2 shows that the unramified Brauer group $\operatorname{Br}^{nr}\big(k(X)\big)$ is an invariant of the variety X with respect to stable birational equivalence. This will be our main tool for establishing stable non-rationality of certain varieties (and in particular their non-rationality in the usual sense).

4.4. Brauer group of a variety

There are several ways to generalize the definition of an unramified Brauer group from fields to varieties (and to schemes as well). The most straightforward one is based on a generalization of central simple algebras. An *Azumaya algebra* over a scheme X is a vector bundle A over X with an algebra structure such that for every schematic point $x \in X$ the fiber $A|_x$ is a central simple algebra over the residue field $k(x)$; see [**Mil80**, Proposition IV.2.1]. Taking a closure of the relation

$$A \sim A \otimes_{\mathcal{O}_X} \operatorname{End}(E),$$

where E is an arbitrary vector bundle over X, we obtain an equivalence relation on Azumaya algebras. Let $\operatorname{Br}(X)$ denote the set of equivalence classes of Azumaya algebras over X; see [**Mil80**, IV.2]. Tensor product of algebras gives a structure of a semigroup on $\operatorname{Br}(X)$. As in the case of fields, there is an isomorphism

$$A \otimes_{\mathcal{O}_X} A^{op} \cong \operatorname{End}(A).$$

This implies that $\operatorname{Br}(X)$ is actually a group.

DEFINITION 4.4.1. The group $\operatorname{Br}(X)$ is called the *Brauer group* of the scheme X.

On the other hand, one can define an unramified Brauer group of an irreducible variety over a field of characteristic zero as follows.

DEFINITION 4.4.2. The *unramified Brauer group* of a normal irreducible variety X over a field k of characteristic zero is the subgroup

$$\mathrm{Br}^{nr}(X) = \bigcap_D \mathrm{Ker}(\mathrm{res}_D)$$

in the group $\mathrm{Br}\left(k(X)\right)$, where D varies over the set of all prime divisors $D \subset X$.

Note that Definition 4.4.2 can also be generalized to the case of non-normal irreducible varieties, but we will not need this.

Finally, one can consider the unramified Brauer group $\mathrm{Br}^{nr}\left(k(X)\right)$ of the function field $k(X)$ on an irreducible variety X over a field of characteristic zero.

Let us describe the relations between the above three groups. There are canonical group homomorphisms

$$\mathrm{Br}(X) \xrightarrow{\phi} \mathrm{Br}^{nr}(X) \xleftarrow{\psi} \mathrm{Br}^{nr}\left(k(X)\right).$$

An important non-trivial fact is that if the variety X is smooth, then ϕ is actually an isomorphism. This can be deduced from the main theorem in [**dJ03**] (which was also proved earlier by Gabber using a different method, but as far as we know his proof was never written down) and the interpretation of the unramified Brauer group $\mathrm{Br}^{nr}(X)$ in terms of étale cohomology generalizing the isomorphism of Exercise 3.1.3(ii); see §A.6 or [**CT95**, §3.4] for details. Moreover, if the variety X is smooth and complete, then ψ is an isomorphism as well; see [**CT95**, Proposition 4.2.3]. In particular, the (unramified) Brauer groups of two stably birational smooth complete varieties over a field of characteristic zero are the same (see the remark after Definition 4.3.3). Stable birational invariance of the group $\mathrm{Br}^{nr}(X)$ can also be proved using just its description via étale cohomology (see §A.5 and §A.6).

It is convenient to use different definitions of Brauer groups of varieties in different situations. For example, given an arbitrary morphism of varieties $f\colon Y \to X$, one can easily define the pull-back homomorphism of Brauer groups

$$f^*\colon \mathrm{Br}(X) \to \mathrm{Br}(Y)$$

by considering pull-backs of vector bundles. If f is an embedding of a point into a smooth curve over a field of characteristic zero, then it is possible to give a simple construction of f^* in terms of unramified Brauer groups. This construction uses the isomorphism from Exercise 4.2.4(i) for the discrete valuation v_Y, and also the homomorphism

$$H^2(G, \mathcal{O}_L^*) \to H^2(G, \lambda^*);$$

see Exercise 4.2.4 for the notation. However, for the arbitrary morphism f, a straightforward description of the pull-back via unramified Brauer groups requires overcoming significant technical difficulties; see [**Ros96**, §12].

The following exercise shows that the residue map on the Brauer group of a function field of a variety has an important property: for every element of the Brauer group, its residue is zero for almost all divisors.

EXERCISE 4.4.3. **Triviality of the residue for almost all divisors**
Let X be a normal irreducible variety over a field k of characteristic zero. Put $K = k(X)$. Take an element $\alpha \in \mathrm{Br}(K)$.

(o) Show that there is a finite morphism of normal irreducible varieties $f\colon Y \to X$ over k such that $\alpha \in \mathrm{Br}(L/K)$ and $K \subset L$ is a finite Galois extension, where $L = k(Y)$. In the remaining parts of this exercise G will denote the Galois group of the extension $K \subset L$.

(i) In the notation of part (o), consider a prime divisor $D \subset X$ and an irreducible component E of the divisor $f^{-1}(D) \subset Y$. Show that the discrete valuation v_E is an extension of the discrete valuation v_D. Prove that all extensions of v_D are of the form v_E for some E constructed like this.

Hint. To prove the first assertion, do a direct computation in the local rings of the divisors D and E. To prove the second assertion, recall that the Galois group G acts on the set of divisors E and apply Exercise 4.1.2(iv).

(ii) Let the subgroup $H \subset G$ consist of the elements $g \in G$ such that $g(E) = E$. Let K_D denote the completion of the field K with respect to the discrete valuation v_D and let L_E denote the completion of L with respect to v_E. Prove that $K_D \subset L_E$ is a Galois extension with Galois group H.

Hint. Use part (i) and Exercise 4.1.2(v).

(iii) Consider the 2-cocycle $\varphi\colon G \times G \to L^*$ that corresponds to the element

$$\alpha \in H^2(G, L^*) \cong \mathrm{Br}(L/K)\,.$$

Define the 2-cocycle ω as the composition of the maps

$$H \times H \to G \times G \xrightarrow{\varphi} L^* \hookrightarrow L_E^*\,.$$

Show that the image of α with respect to the natural map

$$\mathrm{Br}(K) \to \mathrm{Br}(K_D)$$

is given by the class of the 2-cocycle ω.

(iv) Prove that for almost all (that is, for all except a finite number of) prime divisors $E \subset Y$ one has

$$v_E\big(\varphi(g_1, g_2)\big) = 0$$

for all $g_1, g_2 \in G$.

Hint. Given a rational function $\Phi \in k(Y)^*$, one has $v_E(\Phi) = 0$ for almost all prime divisors $E \subset Y$.

(v) Prove that for almost all prime divisors $D \subset X$ one has

$$\mathrm{res}_D(\alpha) = 0\,,$$

where res_D is the residue associated with the discrete valuation defined by the divisor D.

Hint. Almost all divisors D are not contained in the branch divisor $\Delta \subset X$ of the morphism $f\colon Y \to X$, which means that the extension $K_D \subset L_E$ is unramified. Now apply parts (iii) and (iv) together with Exercise 4.2.1(iii).

Exercise 4.4.3 implies that for a (normal) irreducible variety X one has an exact sequence (cf. Exercise 4.3.2)

$$0 \to \mathrm{Br}^{nr}(X) \to \mathrm{Br}\left(k(X)\right) \overset{\oplus_D \mathrm{res}_D}{\longrightarrow} \bigoplus_{D \subset X} \mathrm{Hom}\left(G_{k(D)}, \mathbb{Q}/\mathbb{Z}\right),$$

where the sum is taken over all prime divisors $D \subset X$.

4.5. Geometric meaning of the residue map

Let B be an integral scheme (for example, B can be the spectrum of an integral domain), let K be the field of rational functions on B, and let X be a variety over K. A *model* of the variety X over B is a scheme $\mathcal{X}$ with a flat morphism $\pi\colon \mathcal{X} \to B$ (see [**Har77**, III.9]) and an isomorphism over K between X and the generic schematic fiber of π. If X is projective, that is, there is a closed embedding $X \subset \mathbb{P}^m$ over K, then its *projective model* is a model $\mathcal{X}$ with a closed embedding $\mathcal{X} \subset \mathbb{P}^n_B$ over B. Note that over the generic point this embedding is not required to coincide with the initial projective embedding $X \subset \mathbb{P}^m$ over K. When the latter holds (so that in particular one has $m = n$), we will say that the projective model $\mathcal{X} \subset \mathbb{P}^n_B$ *agrees with the projective embedding* of the variety X. If X is smooth; we will say that $\mathcal{X}$ is a *smooth model* provided that the morphism $\mathcal{X} \to B$ is smooth; see [**Har77**, III.10].

Let R be a Dedekind domain, that is, an integrally closed Noetherian domain such that all its non-zero prime ideals are maximal. For example, R can be a discrete valuation ring, a function ring on a regular affine curve, or a ring of integers in a global field of characteristic zero; see §10.1. Put $B = \mathrm{Spec}(R)$. Let $X \subset \mathbb{P}^n$ be a projective variety over K. Then a projective model $\mathcal{X}$ can be given explicitly as a subscheme in the projective space $\mathbb{P}^n_R$ with homogeneous coordinates $T_0, \ldots, T_n$ as follows. Let I be the ideal in $R[T_0, \ldots, T_n]$ that consists of all homogeneous polynomials that vanish on X. This ideal defines the subscheme $\mathcal{X} \subset \mathbb{P}^n_R$. In other words, $\mathcal{X}$ is given by equations corresponding to the generators of the ideal I. If X is geometrically irreducible, smoothness of $\mathcal{X}$ is equivalent to the fact that for any non-zero prime ideal $\mathfrak{p} \subset R$ the reductions of the equations of $\mathcal{X}$ modulo $\mathfrak{p}$ define a smooth variety over the field $R/\mathfrak{p}$.

Now let R be a valuation ring $\mathcal{O}_K$ in a field K that is complete with respect to a discrete valuation v with perfect residue field κ. Suppose that $X \subset \mathbb{P}^n$ is a hypersurface of degree d over K. Let $\mathcal{X} \subset \mathbb{P}^n_{\mathcal{O}_K}$ be its projective model that agrees with the projective embedding of the hypersurface X over K. Then $\mathcal{X}$ is given in $\mathbb{P}^n_{\mathcal{O}_K}$ by a homogeneous equation $F = 0$ of degree d such that all coefficients of F are from $\mathcal{O}_K$, at least one of them has a non-zero valuation, and the hypersurface given by $F = 0$ over K is projectively equivalent to X. This can be deduced from the facts that the scheme $\mathcal{X}$ is flat over $\mathcal{O}_K$ and that every divisor on the scheme $\mathbb{P}^n_{\mathcal{O}_K}$ can be defined by one equation.

The geometric meaning of the residue map can be formulated as follows: a Severi–Brauer variety X over K has a smooth projective model over $\mathcal{O}_K$ if and only if $\mathrm{res}\left(b(X)\right) = 0$ (see Definition 3.3.3). Below we will prove this fact in a series of exercises. Also, in Exercise 4.5.4 we treat the special case of conics by elementary methods.

EXERCISE 4.5.1. **Triviality of residue**

Let X be a Severi–Brauer variety over K that has a smooth projective model $\mathcal{X}$ over $\mathcal{O}_K$.

(i) Show that the scheme $\mathcal{X}$ is regular, its closed fiber is an irreducible principal divisor, and the restriction map

$$\mathrm{Pic}(\mathcal{X}) \to \mathrm{Pic}(X)$$

is an isomorphism.

(ii) Let $K \subset L$ be an unramified finite Galois extension with Galois group G such that the variety $Y = X_L$ is isomorphic to a projective space over L. Suppose that a divisor $H \subset Y$ corresponds to a hyperplane in the projective space under this isomorphism. Put

$$\mathcal{Y} = \mathrm{Spec}(\mathcal{O}_L) \times_{\mathrm{Spec}(\mathcal{O}_K)} \mathcal{X}\,.$$

Let $\mathcal{H} \subset \mathcal{Y}$ be the Zariski closure of the divisor H in $Y \subset \mathcal{Y}$. The group G acts naturally on the scheme $\mathcal{Y}$. Prove that for every element $g \in G$ there is a rational function f_g on the scheme $\mathcal{Y}$ whose divisor equals $g(\mathcal{H}) - \mathcal{H}$.

Hint. The scheme $\mathcal{Y}$ enjoys all the properties of the scheme $\mathcal{X}$ listed in part (i).

(iii) Prove that $\mathrm{res}\big(b(X)\big) = 0$.

Hint. Apply the construction from Exercise 3.3.5(iv) to the divisor $\mathcal{H}$ and the rational functions f_g on the scheme $\mathcal{Y}$ from part (ii). Since the group of invertible functions on the scheme $\mathcal{Y}$ is $\mathcal{O}_L^*$, this construction gives a class from $H^2(G, \mathcal{O}_L^*)$. Finally, apply Exercise 3.3.6(ii).

EXERCISE 4.5.2. **Cohomology with integral coefficients**

Let $K \subset L$ be an unramified finite Galois extension with Galois group G and let λ be the residue field of the valuation ring $\mathcal{O}_L$ of the field L.

(i) Put $\Gamma = \mathrm{GL}_n(\mathcal{O}_L)$ and consider a decreasing filtration

$$\Gamma^i = I + \mathrm{M}_n\left(\mathfrak{m}_L^i\right), \quad i \geqslant 1\,,$$

where I denotes the identity $n \times n$ matrix and $\mathfrak{m}_L$ is the maximal ideal in $\mathcal{O}_L$. Check that this filtration satisfies all the assumptions of Exercise 2.7.3(ii).

Hint. Construct isomorphisms

$$\Gamma^i/\Gamma^{i+1} \cong \mathrm{M}_n(\lambda)\,, \quad i \geqslant 1\,,$$

where matrices are considered with the additive group law, and use Exercise 2.2.4(iii).

(ii) Show that the image of the filtration from part (i) under the homomorphism

$$\mathrm{GL}_n(\mathcal{O}_L) \to \mathrm{PGL}_n(\mathcal{O}_L)$$

also satisfies all the assumptions of Exercise 2.7.3(ii).

Hint. The corresponding adjoint quotients are isomorphic to the quotient of the group $\mathrm{M}_n(\lambda)$ by scalar matrices.

(iii) Prove that the natural maps

$$H^1\big(G, \mathrm{PGL}_n(\mathcal{O}_L)\big) \to H^1\big(G, \mathrm{PGL}_n(\lambda)\big)$$

are bijections for all $n \geqslant 1$; cf. Exercise 4.2.4(ii).

Hint. Use part (ii) and Exercise 2.7.3(ii).

(iv) Show that there is a canonical bijection between the set $H^1\big(G, \mathrm{PGL}_n(\mathcal{O}_L)\big)$ and the set of isomorphism classes of Azumaya algebras of rank n^2 (see §4.4) over $\mathcal{O}_K$ that are isomorphic to a matrix algebra over $\mathcal{O}_L$.

Hint. Consider a fibered category $\mathcal{M}$ over unramified finite extensions $K \subset E$ such that $\mathcal{M}(E)$ is the category of Azumaya algebras over $\mathcal{O}_E$. Apply the descent theory of §2.1 and §2.2 to this category.

Actually, one can show that every Azumaya algebra over $\mathcal{O}_K$ is isomorphic to a matrix algebra over $\mathcal{O}_E$, where $K \subset E$ is a suitable unramified finite extension.

(v) Choose an element $\alpha \in H^2(G, \mathcal{O}_L^*)$, and denote by $\alpha_\kappa \in H^2(G, \lambda^*)$ its image with respect to the natural map

$$H^2(G, \mathcal{O}_L^*) \to H^2(G, \lambda^*)\,.$$

Choose a finite-dimensional central simple algebra $\mathcal{A}_\kappa$ over κ such that

$$[\mathcal{A}_\kappa] = \alpha_\kappa \in \mathrm{Br}(\kappa)\,.$$

Show that there is a unique (up to isomorphism) Azumaya algebra $\mathcal{A}$ over $\mathcal{O}_K$ whose reduction modulo $\mathfrak{m}_K$ is isomorphic to $\mathcal{A}_\kappa$. Moreover, one has

$$[\mathcal{A}_K] = \alpha_K$$

in $\mathrm{Br}(K)$, where

$$\mathcal{A}_K = K \otimes_{\mathcal{O}_K} \mathcal{A}$$

and α_K is the image of α under the natural map

$$H^2(G, \mathcal{O}_L^*) \to H^2(G, L^*)\,.$$

Hint. Use parts (iii) and (iv).

EXERCISE 4.5.3. **Construction of a smooth model**

(i) Let X be a Severi–Brauer variety over K. Suppose that there is an Azumaya algebra $\mathcal{A}$ over $\mathcal{O}_K$ such that the algebra

$$\mathcal{A}_K = K \otimes_{\mathcal{O}_K} \mathcal{A}$$

corresponds to the Severi–Brauer variety X as described in the beginning of §3.3. Prove that X has a smooth projective model over $\mathcal{O}_K$.

Hint. Apply a construction similar to the one from the beginning of §3.3 to the algebra $\mathcal{A}$ over $\mathcal{O}_K$. Specifically, consider the Grassmannian $\mathrm{G}(\mathcal{A}, n)$ over $\mathcal{O}_K$ and a subscheme $\mathcal{X}$ over $\mathcal{O}_K$ therein that parameterizes right ideals in $\mathcal{A}$. The generic fiber $\mathcal{X}_K$ is isomorphic to X, and the closed fiber is the Severi–Brauer variety corresponding to the finite-dimensional central simple algebra

$$\mathcal{A}_\kappa = \kappa \otimes_{\mathcal{O}_K} \mathcal{A}$$

over κ. Therefore, $\mathcal{X}$ is a smooth projective model of the variety X over $\mathcal{O}_K$.

(ii) Suppose that there exists an Azumaya algebra $\mathcal{D}$ over $\mathcal{O}_K$ such that $D = \mathcal{D}_K$ is a division algebra. Show that for every finite-dimensional central simple algebra A over K with $b(A) = b(D)$, there is an Azumaya algebra $\mathcal{A}$ over $\mathcal{O}_K$ such that $A \cong \mathcal{A}_K$.

Hint. General theory of finite-dimensional central simple algebras, as outlined in §1.1, implies that $A \cong \mathrm{M}_p(D)$ for some positive integer p.

(iii) Let $\mathcal{D}$ be an Azumaya algebra over $\mathcal{O}_K$ such that the algebra $\mathcal{D}_\kappa$ over κ is a division algebra. Prove that the algebra $\mathcal{D}_K$ over K is a division algebra as well.

Hint. Consider the hypersurface $\mathcal{H}$ in $\mathbb{P}^n_{\mathcal{O}_K}$ defined by the reduced norm of the algebra $\mathcal{A}$; see Exercise 3.1.12. Since $\mathcal{D}_\kappa$ is a division algebra, the hypersurface $\mathcal{H}_\kappa$ has no κ-points by Exercise 3.1.12(v). Hence $\mathcal{H}$ has no $\mathcal{O}_K$-points. Since $\mathcal{H}$ is projective over $\mathcal{O}_K$, this means that the hypersurface $\mathcal{H}_K$ over K has no K-points. Using Exercise 3.1.12(v) once again, we see that $\mathcal{D}_K$ is a division algebra.

(iv) Let X be a Severi–Brauer variety over K such that $\mathrm{res}\big(b(X)\big) = 0$. Prove that X has a smooth model over $\mathcal{O}_K$.

Hint. Let $K \subset L$ be an unramified Galois extension with Galois group G such that

$$b(X) \in \mathrm{Br}(L/K)\,.$$

By Exercise 4.2.4(i) the assumption on the residue implies the existence of an element $\alpha \in H^2(G, \mathcal{O}_L^*)$ such that its image in $H^2(G, L^*)$ equals $b(X)$. Consider its reduction

$$\alpha_\kappa \in H^2(G, \lambda^*)\,.$$

Let $\mathcal{D}_\kappa$ be a division algebra over κ whose class in $\mathrm{Br}(\kappa)$ equals α_κ. Applying Exercise 4.5.2(v), we obtain an Azumaya algebra $\mathcal{D}$ over $\mathcal{O}_K$. It remains to use parts (i), (ii), and (iii).

Now we will consider in detail the case of conics. In this case the equivalence between triviality of the residue and existence of a smooth model can be established by elementary methods. We assume for simplicity that the characteristic of the field κ is different from 2 (and thus the characteristic of the field K is different from 2 as well). Let $X \subset \mathbb{P}^2$ be a smooth conic over K. By a *good* model of X over $\mathcal{O}_K$ we will mean a projective model $\mathcal{X} \subset \mathbb{P}^2_{\mathcal{O}_K}$ that agrees with the embedding $X \subset \mathbb{P}^2_K$ such that the scheme $\mathcal{X}$ is regular and the closed fiber

$$\mathcal{X}_\kappa = \mathrm{Spec}(\kappa) \times_{\mathrm{Spec}(\mathcal{O}_K)} \mathcal{X}$$

is either smooth or a pair of different lines in $\mathbb{P}^2$ (cf. Exercise 6.3.1).

Exercise 4.5.4. **Models of conics**

(i) Prove that every good model of the smooth conic X is given by an equation of the form

$$x^2 - ay^2 - bz^2 = 0\,, \tag{4.1}$$

where $a, b \in \mathcal{O}_K$ and at least one of the coefficients a and b is invertible in $\mathcal{O}_K$, that is, has a zero valuation, while the valuation of the other coefficient is equal to either 0 or 1.

Hint. Since 2 is invertible in $\mathcal{O}_K$, the equation of the model $\mathcal{X}$ can be diagonalized. The assumption on the closed fiber implies that at least two coefficients in the diagonal equation are invertible. The assumption about the regularity of $\mathcal{X}$ implies that the valuation of the remaining coefficient cannot exceed 1.

(ii) Prove that every smooth conic X over K has a good model over $\mathcal{O}_K$.

Hint. An equation of the conic can be transformed to the form (4.1) by multiplying it by a suitable power of a uniformizer of $\mathcal{O}_K$ and making a diagonal change of variables.

(iii) Choose a good model $\mathcal{X}$ of the conic X. Check that there is a G_κ-equivariant bijection between the set of irreducible components of the variety $\mathcal{X}_{\bar{\kappa}}$ and the set $\left\{\pm\sqrt{(a,b)_v}\right\}$ (see Definition 4.2.2), where a and b are defined as in part (i). If the model $\mathcal{X}$ is smooth, then we let the first action be the trivial action on the set of two elements by definition.

Hint. Using part (i), consider the cases of a smooth and a singular fiber.

In particular, it follows from Exercises 3.3.2(i) and 4.2.3(iv) that the action of the group G_κ on the set of irreducible components of the variety $\mathcal{X}_{\bar{\kappa}}$ does not depend on the choice of a good model.

(iv) Suppose that the action of the Galois group G_κ on the set of irreducible components of the variety $\mathcal{X}_{\bar{\kappa}}$ is trivial. Show that X has a smooth projective model over $\mathcal{O}_K$.

Hint. In this case the conic X has a point over K, and thus it is projectively equivalent to any other conic with a point, such as the conic given by the equation

$$x^2 - y^2 - z^2 = 0.$$

A more geometric approach is as follows. In the case of a singular closed fiber, by our assumption each of the irreducible components of the closed fiber $X_{\bar{\kappa}}$ is defined over κ. Thus one can contract any of them on the scheme $\mathcal{X}$ and get a smooth model over $\mathcal{O}_K$.

(v) Show that for any smooth projective model $\mathcal{X}$ of the conic X there is a closed embedding $\mathcal{X} \subset \mathbb{P}^2_{\mathcal{O}_K}$ that agrees with the embedding $X \subset \mathbb{P}^2_K$.

Hint. The closed fiber of a smooth model $\mathcal{X}$ is a form of $\mathbb{P}^1$ because the relative canonical sheaf has degree -2 on every fiber. Embed the scheme $\mathcal{X}$ into $\mathbb{P}^2_{\mathcal{O}_K}$ via the anticanonical sheaf.

(vi) Prove that the conic X has a smooth projective model if and only if $\operatorname{res}\big(b(X)\big) = 0$.

Hint. Use the previous parts of this exercise together with Exercise 4.2.3(iv).

A detailed treatment of degenerations of models of Severi–Brauer varieties can be found in [**Art82**]. In particular, it is proved there that every Severi–Brauer variety X over K has a projective model $\mathcal{X}$ such that it is a regular scheme and the closed fiber $\mathcal{X}_\kappa$ is irreducible. The algebraic closure λ of the field κ in the function field $\kappa(\mathcal{X}_\kappa)$ (cf. Exercise 2.5.1(ii)) is a cyclic extension of κ given by the residue

$$\mathrm{res}\big(b(X)\big) \in \mathrm{Hom}(G_\kappa, \mathbb{Q}/\mathbb{Z})\,.$$

Moreover, the variety $\mathcal{X}_{\bar{\kappa}}$ is a union of r copies of one smooth projective rational variety that transversally meet each other, where r is equal to the order of the residue $\mathrm{res}\big(b(X)\big)$ in the group $\mathrm{Hom}(G_\kappa, \mathbb{Q}/\mathbb{Z})$; a generator of the Galois group

$$\mathrm{Gal}(\lambda/\kappa) \cong \mathbb{Z}/r\mathbb{Z}$$

acts by a cyclic permutation on the set of irreducible components of the variety $\mathcal{X}_{\bar{\kappa}}$. This is a direct generalization of the case of conics considered in Exercise 4.5.4.

4.6. Further reading

The reader can find foundational material on Brauer groups of varieties in [**Gro95a**], [**Gro95b**], and [**Gro68**] (see also [**DF84**] for amendments of some inaccuracies in the latter paper). A good introduction to unramified Brauer groups and more general stable birational invariants is contained in [**CTS07**].

Part III

Applications to rationality problems

CHAPTER 5

Example of a Unirational Non-rational Variety

In this chapter we produce an example of a variety $X = V/G$, where G is a finite group and V is its representation defined over an algebraically closed field k of characteristic zero, such that non-rationality of V can be proved using the methods introduced in the previous chapters. Examples of this kind first appeared in the papers by D. Saltman [**Sal84**] and F. A. Bogomolov [**Bog87**], but our presentation follows a simplified approach due to I. R. Shafarevich [**Sha90**].

Note that one can obtain a similar example when k has positive characteristic l different from p, where p is the prime in §5.2 below. This is because all facts about the unramified Brauer group that we use in the characteristic-zero case remain valid for the subgroup in the Brauer group that consists of all elements of order coprime to l (in particular, for the p-torsion subgroup), which can be shown using étale cohomology; see [**CT95**, §§3.3, 3.4, 4.2].

5.1. Geometric data

Let G be a finite subgroup in the automorphism group of a smooth irreducible variety V defined over a field k. Put $L = k(V)$ and $K = k(V)^G$. The field extension $K \subset L$ is a Galois extension with Galois group G.

EXERCISE 5.1.1. **Relative Brauer group**
Suppose that $\mathrm{Pic}(V) = 0$ and $k[V]^* = k^*$. Prove that there is a natural embedding

$$H^2(G, k^*) \hookrightarrow \mathrm{Br}(L/K)\,.$$

Hint. Use the exact sequence of G-modules

$$1 \to k^* \to k(V)^* \to \mathrm{Div}(V) \to 0$$

and Exercise 2.3.10(i).

Now suppose that the field k is algebraically closed. Denote by μ the group of roots of unity in k.

EXERCISE 5.1.2. **Replacing k^* by μ**
Show that

$$H^i(G, \mu) \cong H^i(G, k^*)$$

for all $i > 0$.

Hint. Use the fact that the group k^*/μ is uniquely divisible and Exercise 1.3.3(i).

In what follows, $Z(\Gamma)$ will denote the center of a group Γ.

Suppose that there is an element $0 \neq \alpha \in H^2(G, \mu)$ with the following property:

(5.1) for every subgroup $H \subset G$ and every central cyclic subgroup $N \subset Z(H)$ in H, the restriction $\alpha|_H$ of the element α to the subgroup H is contained in the image of the natural map $H^2(H/N, \mu) \to H^2(H, \mu)$.

Let $\bar{\alpha} \in \mathrm{Br}(K)$ be the image of α under the embedding

$$H^2(G, \mu) \hookrightarrow \mathrm{Br}(K),$$

that is, under the composition of the isomorphism

$$H^2(G, \mu) \cong H^2(G, k^*)$$

from Exercise 5.1.2, the embedding

$$H^2(G, k^*) \hookrightarrow \mathrm{Br}(L/K)$$

from Exercise 5.1.1, and the natural embedding $\mathrm{Br}(L/K) \hookrightarrow \mathrm{Br}(K)$.

From now on we assume that the field k is of characteristic zero, so that we have a well-defined residue for elements of Brauer groups (see Exercise 4.2.1 and §4.3).

EXERCISE 5.1.3. **Quotient by a finite group action**

(i) With the above notation, prove that for any discrete valuation v on the field K that is trivial on k, one has $\mathrm{res}_v(\bar{\alpha}) = 0$.

Hint. Let w be an arbitrary extension of the valuation v to L. By Exercise 4.1.2(v), the extension of the corresponding complete fields $K_v \subset L_w$ is Galois and its Galois group H is canonically embedded into G. The image of α with respect to the map $\mathrm{Br}(K) \to \mathrm{Br}(K_v)$ is given by the restriction $\alpha|_H \in H^2(H, \mu)$. By Exercise 4.1.1(iv) and condition (5.1), the element $\alpha|_H$ is contained in the image of the natural map

$$H^2(H^{nr}, \mu) \to H^2(H, \mu).$$

Now use Exercise 4.2.7.

Therefore, one has $\bar{\alpha} \in \mathrm{Br}^{nr}(K)$.

(ii) Use part (i) to show that the variety V/G is not stably rational.

Hint. Apply Exercise 4.3.2.

In particular, if V is a faithful representation of the group G, then the quotient V/G is a unirational but not stably rational variety.

5.2. Construction of a group

Since the characteristic of the field k is zero, a (non-canonical) consistent choice of primitive roots of unity defines an isomorphism of G-modules $\mathbb{Q}/\mathbb{Z} \cong \mu$. We are going to construct a group G such that there is an element

$$0 \neq \alpha \in H^2(G, \mathbb{Q}/\mathbb{Z}) \cong H^2(G, \mu)$$

that satisfies condition (5.1) above. Let $p \neq 2$ a prime number and $W \cong \mathbb{F}_p^4$ a four-dimensional vector space over the finite field $\mathbb{F}_p$. Put

$$W' = \Lambda^2(W) \cong \mathbb{F}_p^6,$$

where Λ^2 denotes the exterior square of a vector space. Let the group $\widetilde{G}$ be a central extension

$$0 \to W' \to \widetilde{G} \to W \to 0$$

such that the corresponding 2-cocycle is the pairing

$$\omega\colon (w_1, w_2) \mapsto w_1 \wedge w_2 \in W',$$

where W' is considered as a trivial W-module (see Exercise 1.1.4(ii)). The group $\widetilde{G}$ has order p^{10}, nilpotency class 2, and exponent p. One has

$$W' = [\widetilde{G}, \widetilde{G}] = Z(\widetilde{G}) \subset \widetilde{G}.$$

All of this is straightforward to check using nothing but the definition of $\widetilde{G}$.

EXERCISE 5.2.1. **Properties of the group $\widetilde{G}$**

(i) Check that ω is indeed a 2-cocycle.

(ii) Show that in the group $\widetilde{G}$ one has

$$[g_1, g_2] = 2w_1 \wedge w_2,$$

where $w_i \in W$ and $g_i \in \widetilde{G}$ is an arbitrary preimage of the element w_i.

(iii) Prove that for any non-trivial subgroup $M \subset W'$ the extension

$$0 \to M \to \widetilde{G} \to \widetilde{G}/M \to 1$$

is non-trivial.

Hint. Choose a non-trivial element $m \in M$ and elements $w_i, v_i \in W$ such that

$$2\sum_i w_i \wedge v_i = m.$$

Then for any preimages $g_i, f_i \in \widetilde{G}$ of the elements w_i and v_i one has

$$\prod_i [g_i, f_i] = m \in \widetilde{G}.$$

On the other hand, if our extension is trivial, then $\widetilde{G}$ contains a subgroup $\widetilde{G}_M \cong \widetilde{G}/M$ and $[\widetilde{G}_M, \widetilde{G}_M] \subset \widetilde{G}_M$. It remains to notice that one can choose $g_i, f_i \in \widetilde{G}_M \subset \widetilde{G}$ to obtain a contradiction.

EXERCISE 5.2.2. **Non-triviality of extensions**

Let Γ be an arbitrary group.

(i) Consider an element

$$\varphi \in \mathrm{Hom}(\Gamma, \mathbb{Q}/\mathbb{Z}) \cong H^1(\Gamma, \mathbb{Q}/\mathbb{Z})$$

and the coboundary map

$$\delta\colon H^1(\Gamma, \mathbb{Q}/\mathbb{Z}) \to H^2(\Gamma, \mathbb{Z}/p\mathbb{Z})$$

that comes from the exact sequence of trivial Γ-modules

$$0 \to \mathbb{Z}/p\mathbb{Z} \to \mathbb{Q}/\mathbb{Z} \xrightarrow{p} \mathbb{Q}/\mathbb{Z} \to 0. \tag{5.2}$$

Prove that the element $\delta(\varphi) \in H^2(\Gamma, \mathbb{Z}/p\mathbb{Z})$ corresponds to the extension

$$0 \to \mathbb{Z}/p\mathbb{Z} \to \widetilde{\Gamma} \to \Gamma \to 1$$

defined as the pull-back of the extension (5.2) with respect to the homomorphism $\varphi\colon \Gamma \to \mathbb{Q}/\mathbb{Z}$.

(ii) Suppose that a non-zero element $\beta \in H^2(\Gamma, \mathbb{Z}/p\mathbb{Z})$ corresponds to an extension

$$0 \to \mathbb{Z}/p\mathbb{Z} \to \widetilde{\Gamma} \to \Gamma \to 1$$

such that the group $\widetilde{\Gamma}$ has exponent p. Show that the image of the element β under the natural map

$$H^2(\Gamma, \mathbb{Z}/p\mathbb{Z}) \to H^2(\Gamma, \mathbb{Q}/\mathbb{Z})$$

is non-trivial.

Hint. Assume that it is trivial and consider a long exact sequence of cohomology groups. Use part (i) to construct an element of $\widetilde{\Gamma}$ whose order exceeds p.

Choose a non-trivial element $z \in W' \subset \widetilde{G}$; until the very last step we will not use any specific properties of z. Consider the subgroup

$$\mathbb{Z}/p\mathbb{Z} \cong \langle z \rangle \subset \widetilde{G}$$

generated by z. Define the group G as $\widetilde{G}/\langle z \rangle$, and consider the element $\beta \in H^2(G, \mathbb{Z}/p\mathbb{Z})$ that corresponds to the extension

$$0 \to \langle z \rangle \to \widetilde{G} \to G \to 1\,.$$

Let $H \subset G$ be an arbitrary subgroup. Consider an element $u \in Z(H)$ and the (central cyclic) subgroup $N \subset H$ generated by u. Let

$$I \subset H^2(H, \mathbb{Z}/p\mathbb{Z})$$

be the image of the natural map

$$H^2(H/N, \mathbb{Z}/p\mathbb{Z}) \to H^2(H, \mathbb{Z}/p\mathbb{Z})\,,$$

and let $\gamma \in H^2(H, \mathbb{Z}/p\mathbb{Z})$ be the restriction of the 2-cocycle β from G to H.

EXERCISE 5.2.3. **Condition (5.1) for the group G**

(o) Consider the central extension

$$0 \to \langle z \rangle \to \widetilde{H} \to H \to 1 \tag{5.3}$$

corresponding to the 2-cocycle γ. Show that γ is contained in I if and only if the extension (5.3) has a (group-theoretic) section over N whose image is a normal subgroup in $\widetilde{H}$.

(i) Prove that

$$Z(G) = Z(\widetilde{G})/\langle z \rangle = W'/\langle z \rangle$$

and there is an exact sequence

$$0 \to Z(G) \to G \to W \to 0\,.$$

(ii) Suppose that $u \in Z(G)$. Prove that $\gamma \in I$.

Hint. Use parts (o) and (i) together with the fact that W' is an abelian group of exponent p.

(iii) Suppose that $u \notin Z(G)$ and the image $v \in W$ of the element u generates the image of the subgroup H with respect to the homomorphism $G \to W$. Prove that $\gamma = 0$.

Hint. In this case the group H and its preimage $\widetilde{H}$ in $\widetilde{G}$ are abelian groups of exponent p, so that $\langle z \rangle \subset \widetilde{H}$ splits as a direct summand.

(iv) Suppose that $u \notin Z(G)$ and $u' \in H$ is an element whose image v' in W is not contained in the subgroup generated by v. Prove that
$$v \wedge v' = lz$$
for some $l \not\equiv 0 \pmod{p}$.

Hint. Use Exercise 5.2.1(ii).

(v) Show that for a suitable choice of z, the situation described in part (iv) does not take place for any H and $u \in Z(H)$.

Hint. Choose the element
$$z \in W' = \Lambda^2 W$$
to be an indecomposable bivector.

(vi) Prove that the element $\beta \in H^2(G, \mathbb{Z}/p\mathbb{Z})$ is non-zero.

Hint. Apply Exercise 5.2.1(iii).

(vii) Show that the image α of the element β under the natural map
$$H^2(G, \mathbb{Z}/p\mathbb{Z}) \to H^2(G, \mathbb{Q}/\mathbb{Z})$$
is non-trivial.

Hint. Apply part (vi) and Exercise 5.2.2(ii).

Using parts (ii), (iii), and (v), prove that for a suitable choice of z the element $\alpha \in H^2(G, \mathbb{Q}/\mathbb{Z})$ satisfies condition (5.1). Therefore, for any faithful representation V of our group G over an algebraically closed field k of characteristic zero, the variety V/G is not stably rational by Exercise 5.1.3.

5.3. Further reading

Given a finite group G and its faithful representation V, it is natural to ask if the quotient V/G is rational (or stably rational). This question is known as *Noether's problem.* Note that the stable birational equivalence class of the quotient V/G depends only on G; this result is sometimes referred to as the *no-name lemma*; see for instance [**CTS07**, Corollary 3.9]. Solving Noether's problem in different cases requires both studying obstructions to rationality and finding explicit rationality constructions.

The unramified Brauer group of the function field of a quotient V/G of a faithful representation V of a finite group G was described in group-theoretic terms in [**Bog87**]. Since then it has often been denoted by $B_0(G)$ and called the *Bogomolov multiplier* of the group G. Note that $B_0(G)$ depends only on G, not on V. The group $B_0(G)$ is the most accessible obstruction to stable rationality of quotients V/G.

Effective methods for computing unramified Brauer groups for fields of invariants $k(V)^G$ in terms of bi-cyclic subgroups of G are described in [**Bog87**] (see also [**CTS07**, §7]). An algorithm to compute $B_0(G)$ for solvable groups was described in [**Mor12b**]. It is known that $B_0(G)$ depends only on a class of isoclinism of the group G; see [**Mor14**], and also [**BB13**, Theorem 3.2] for a more general (and more geometric) result. This can often reduce the required computations.

Note that in many important cases, including simple groups, $B_0(G)$ vanishes (and thus does not provide any obstruction to rationality of the quotient); see for instance [**Kun10**], [**KK14**], or [**JM15**] (cf. also [**BP11**]). However, p-groups provide a large and interesting class of groups where $B_0(G)$ is often useful for detecting non-rationality of quotients (here p is a prime number as usual). Moreover, to some extent the whole theory of Bogomolov multipliers is defined by p-groups; see [**Bog87**, §4].

The example considered in this chapter is not optimal; that is, the group described in §5.2 has order p^9. Actually, one can provide a similar example with a group of order p^6, and also for $p = 2$ (see [**Bog87**]). Moreover, for $p > 2$ one can do the same for a group of order p^5; see [**HK11**], [**Mor12b**], [**Mor12a**], and [**HKK13**]. On the other hand, for a group G of order p^k with $k \leqslant 4$, all quotients V/G of faithful representations defined over an algebraically closed field of characteristic zero are rational (see [**CK01**] and [**Kan09**]).

In some instances 2- and 3-groups may behave in a different way from p-groups with $p \geqslant 5$. Noether's problem for groups of order 2^k, $k \leqslant 6$, was completely solved; see [**CHKP08**], [**CHKK10**], and [**Kor15**]. Also, there are partial results for groups of order $128 = 2^8$; see [**JM14**] and [**Hos16**]. As for 3-groups, at the moment everything is known up to order $243 = 3^5$; see [**CHHK15**] and [**HKY16**].

The unramified Brauer group is not the only known obstruction to stable rationality: there are higher unramified cohomology groups generalizing it, with the second cohomology group corresponding to the unramified Brauer group. One can read about them in [**CTO89**] and [**Pey93**]; these papers also contain examples of obstructions to stable rationality based on higher cohomology groups. The third unramified cohomology group is particularly well studied. There are examples of quotient varieties $X = V/G$, where V is a representation of a finite group G over an algebraically closed field of characteristic zero whose stable rationality is not obstructed by the unramified Brauer group, but is obstructed by the third unramified cohomology; see [**Pey08**], [**HKY16**], and [**HKY16**] (cf. also [**Aso13**] for examples where non-rationality is detected only by sufficiently high unramified cohomology groups; however, these examples have nothing to do with quotients by finite groups).

Another thing to mention is that over some non-algebraically closed fields (in particular, over $\mathbb{Q}$) there exist non-stably rational varieties of the form V/Γ, where V is a representation of a cyclic group $\Gamma \cong \mathbb{Z}/n\mathbb{Z}$; see [**Swa69**], [**Len74**], and [**Pla17**].

In the cases where the solution of Noether's problem is positive, one often encounters interesting rationality constructions. Such constructions are known for some subgroups of symmetric groups and their central extensions ([**SB89**], [**Mae89**], [**Pla09**], [**HK10**], [**Zho15**], [**KWZ15**], [**KW14**], [**KZ12**]), for some semi-direct products (see [**KZ17**]), and for certain classes of p-groups ([**Kan06**], [**Mic14**]). Several general reduction theorems for rationality problem can be found in [**KP09**].

Instead of quotients of linear representations, one can consider quotients of their projectivizations (keeping in mind that these are stably birational; see [**Pro10**, Proposition 1.2]). Of course, all quotients of $\mathbb{P}^1$ and $\mathbb{P}^2$ over an algebraically closed field (of characteristic zero) are rational. Actually, all quotients of $\mathbb{P}^2$ are rational over an arbitrary field of characteristic zero as well; see [**Tre14**]. A survey of rationality constructions for quotients of $\mathbb{P}^3$ can be found in [**Pro10**].

For other relevant results and further details we refer the reader to the surveys [**CTS07**] and [**BT17**].

CHAPTER 6

Arithmetic of Two-dimensional Quadrics

6.1. Invariants of quadrics

Let K be an arbitrary field of characteristic different from 2. Consider a four-dimensional vector space V over K and a non-degenerate quadratic form q on V. Let $Q \subset \mathbb{P}(V) \cong \mathbb{P}^3$ be a two-dimensional quadric defined by the form q, that is, Q is given by the equation $q = 0$.

As before, for a field extension $K \subset L$ we will denote the scalar extension of any object X from K to L by X_L. In particular, we can consider a quadratic form q_L on the vector space $V_L = L \otimes_K V$ over L and a quadric Q_L over L.

DEFINITION 6.1.1. Choose a basis of V over K and consider the determinant of the matrix corresponding to the quadratic form q. Note that its class in the quotient group $K^*/(K^*)^2$ does not depend on the choice of basis. Also it does not depend on the choice of q provided that the quadric Q remains the same, because the dimension of V is even. This class in $K^*/(K^*)^2$ is called the *discriminant* of the quadric Q and is denoted by $\mathrm{d}(Q)$.

One can easily see from Definition 6.1.1 that the discriminant is well-defined for any smooth even-dimensional quadric.

EXERCISE 6.1.2. **Quadrics with a trivial discriminant**
Suppose that $\mathrm{d}(Q) = 1$.

(i) Show that Q is a *Pfister quadric*, that is, it is given by the equation

$$x^2 - ay^2 - bz^2 + abt^2 = 0$$

for some $a, b \in K^*$.

(ii) Let $C_0 \subset \mathbb{P}^2$ be a conic over the field K given by the equation

$$u^2 - av^2 - bw^2 = 0\,,$$

where $a, b \in K^*$ are defined by the quadric Q as explained in part (i). In other words, C_0 is the section of the quadric Q by the plane $t = 0$. Prove that there is a projection $\pi_0 \colon Q \to C_0$ defined over the field K whose fibers are lines in the embedding $Q \subset \mathbb{P}^3$ (cf. Exercise 3.1.5(ii)) and that the restriction of π_0 to the plane section $C_0 \subset Q$ is the identity map.

Hint. Let π_0 map the point $(x : y : z : t)$ to the point $(u : v : w)$, where $w = z^2 - at^2$ and

$$u + \sqrt{a}\, v = (x + \sqrt{a}\, y)(z + \sqrt{a}\, t)\,.$$

To check that this definition makes sense, one can use the fact that the norm map

$$x + \sqrt{a}\, y \mapsto x^2 - ay^2$$

is a homomorphism of multiplicative groups. A fiber at a point $(u : v : w)$ of the conic C_0 is a line on Q that passes through the points $(u : v : w : 0)$ and $(av : u : 0 : -w)$; cf. Exercise 3.3.2(i).

(iii) Consider an involution $\iota_0 \colon Q \to Q$ given by the formula

$$(x : y : z : t) \mapsto (x : y : z : -t) .$$

Check that the composition $\pi_0' = \pi_0 \circ \iota_0$ has all the properties of the projection π_0 listed in part (ii). Therefore, the quadric Q contains two families of lines defined over K. Using the fact that Q does not contain other families of lines (even over the algebraic closure of K), show that for a given conic $C_0 \subset Q$ the projections π_0 and π_0' are the only morphisms with these properties. In particular, a choice of one of the two families of lines on Q (or, which is the same, on $Q_{\bar{K}}$) is equivalent to a choice of one of the two morphisms π_0 and π_0'.

(iv) Fix one of the two families of lines on Q. Prove that for every smooth plane section C of Q this gives an isomorphism $C \cong C_0$.

Hint. By part (iii) we have chosen one of the two projections π_0 and π_0'. Its restriction to C gives the required isomorphism.

(v) Let C be a smooth plane section of the quadric Q. Prove that there are exactly two projections $\pi, \pi' \colon Q \to C$ that restrict to the identity map on C and whose fibers are lines (cf. part (iii)).

(vi) Fix one of the two families of lines on Q. Prove that this gives an isomorphism $Q \cong C \times C$ of surfaces over K, where C is an arbitrary smooth plane section of the quadric Q.

Hint. Define projections $Q \to C$ as in part (v).

(vii) Give an explicit proof of the fact that the fibers of the projection $C \times C \to C$ are isomorphic to $\mathbb{P}^1$.

Hint. The diagonal $C \to C \times C$ gives a section of a family of Severi–Brauer varieties $C \times C \to C$. Now one can apply Exercise 2.2.8(i).

Actually, one can show that the $\mathbb{P}^1$-bundle $C \times C \to C$ is the projectivization $\mathbb{P}(V) \to C$ of a rank-2 vector bundle V on C defined as follows:

$$V = \operatorname{Ker}\left(H^0(C, T_C) \otimes \mathcal{O}_C \to T_C\right),$$

where T_C is the tangent bundle on C.

(viii) Prove that for an arbitrary smooth quadric $Q \subset \mathbb{P}^3$ over K, the following conditions are equivalent: the discriminant of Q is trivial; Q is a Pfister quadric; there is an isomorphism $Q \cong C \times C$ of surfaces over K, where C is some smooth conic over K.

Hint. To check that the first and second conditions are equivalent, use part (i). To check that the second and third are equivalent, use part (vi).

The main assertions from Exercise 6.1.2 can be interpreted in terms of quaternion algebras. Specifically, consider the quaternion algebra $A(a, b)$ (see Exercise 3.1.5), where $a, b \in K^*$. The hypersurface in $\mathbb{P}(A)$ given by the equation

$$\operatorname{Nrd}(\alpha) = 0, \quad \alpha \in A$$

(see Exercise 3.1.12), is the quadric Q. The conic C_0 corresponds to the algebra A via Exercise 3.3.2(i), and two families of lines on Q with base C_0 are families of right and left ideals in A (the morphism π_0 corresponds to right ideals, and π_0' corresponds to left ideals).

Exercise 6.1.2(iv) implies that if the discriminant $\mathrm{d}(Q)$ is trivial, then there is a well-defined conic C over the field K that is isomorphic to (every) smooth plane section of the quadric Q.

DEFINITION 6.1.3. In this case the class $b(C)$ of the conic C in the Brauer group $\mathrm{Br}(K)$ (see Definition 3.3.3) is called the *Clifford invariant* of the quadric Q and is denoted by $\mathrm{cl}(Q)$.

The Clifford invariant is actually well-defined for an arbitrary smooth even-dimensional quadric with trivial discriminant. Therefore, we have a sequence of invariants of quadrics (parity of dimension, discriminant, and Clifford invariant) where each of them is well-defined if all previous invariants vanish. They take values in the groups $\mathbb{Z}/2\mathbb{Z}$, $K^*/(K^*)^2$, and $\mathrm{Br}(K)_2$, respectively. By Exercises 2.4.2(ii) and 3.1.10(iii), these groups are isomorphic to Galois cohomology groups $H^n(G_K, \mathbb{Z}/2\mathbb{Z})$ with $n = 0, 1, 2$.

There is a generalization of these invariants for arbitrary values of n. For $n = 3$ such an invariant was constructed by J. Arason in [**Ara75**]. For some particular quadrics the invariants can be constructed explicitly; see [**EKM08**, §18]. In the general case the construction is based on two Milnor conjectures that are now already proved. One of them relates quadratic forms over a field K to Milnor K-groups $\mathrm{K}_n^M(K)$ (see Definition 3.1.7 for $n = 2$); it was proved by D. Orlov, A. Vishik, and V. Voevodsky in [**OVV07**]. Another conjecture states that the natural homomorphism

$$\mathrm{K}_n^M(K)/2 \to H^n(G_K, \mathbb{Z}/2\mathbb{Z})$$

is an isomorphism for every positive integer n; it was proved by V. Voevodsky in [**Voe03**]. Note that both this assertion and the Merkurjev–Suslin theorem (see the discussion after Exercise 3.1.9) are particular cases of a more general Bloch–Kato conjecture recently proved by V. Voevodsky and M. Rost.

EXERCISE 6.1.4. **Non-triviality of the Clifford invariant**

(i) Let $F = K(\sqrt{a})$ be an arbitrary quadratic extension of the field K. Suppose that the quadric Q_F contains a line and that $Q(K) = \varnothing$. Show that the quadratic form q is a tensor product of two quadratic forms on two-dimensional vector spaces. In particular, the discriminant $\mathrm{d}(Q)$ is trivial.

Hint. Let $U \subset V_F$ be a two-dimensional subspace over F corresponding to a line on Q_F. Then U is isotropic with respect to the form q_F, and one has $U \cap \sigma(U) = 0$, where σ denotes the non-trivial automorphism of the field F over K. Show that U has a basis $\{e, f\}$ over F such that

$$b\big(e, \sigma(f)\big) = 0,$$

where b is the symmetric bilinear form on V_F associated with q_F. Finally, consider the basis

$$\big\{e + \sigma(e),\ (e - \sigma(e))/\sqrt{a},\ f + \sigma(f),\ (f - \sigma(f))/\sqrt{a}\big\}$$

in V over K and the corresponding Gram matrix.

Note that in a similar way one can prove a generalization of this fact to the case of quadratic forms in a larger number of variables. Let q be a non-degenerate quadratic form in $2n+2$ variables, and let Q be the corresponding quadric in $\mathbb{P}^{2n+1}$. Suppose that the quadric Q_F contains a subspace $\mathbb{P}^n$ and that $Q(K) = \varnothing$. Then the form q is (tensor) divisible by a quadratic form on a two-dimensional vector space given by a diagonal matrix $\mathrm{diag}(1, -a)$.

(ii) Consider the field

$$L = K\big(\sqrt{\mathrm{d}(Q)}\big).$$

Note that L coincides with K if $\mathrm{d}(Q)$ is trivial, and L is a quadratic extension of K if $\mathrm{d}(Q)$ is non-trivial. Let the conic C over L be a smooth plane section of the quadric Q_L. Prove that $C(L) \neq \varnothing$ if and only if $Q(K) \neq \varnothing$.

Hint. If the discriminant is trivial, use Exercise 6.1.2(ii) and (iv). If the discriminant is non-trivial and $C(L) \neq \varnothing$, apply Exercise 6.1.2(v) and prove that the quadric Q_L contains a line. Then use part (i). To prove the converse implication, note that if $Q(K) \neq \varnothing$, then $Q_L(L) \neq \varnothing$, so that everything is reduced to the case of a trivial discriminant.

(iii) In the notation of part (ii), prove that the equality $\mathrm{cl}(Q_L) = 0$ holds in the group $\mathrm{Br}(L)$ if and only if $Q(K) \neq \varnothing$.

Hint. Use part (ii) and Exercise 3.3.4.

6.2. Geometric meaning of invariants of quadrics

EXERCISE 6.2.1. **Geometric meaning of the discriminant**

(i) Consider a line l on $Q_{K^{sep}}$ and an element g of the Galois group G_K. Prove that the intersection $g(l) \cap l$ is a point, that is, l and $g(l)$ are contained in different families of lines on $Q_{K^{sep}}$, if and only if

$$g\big(\sqrt{\mathrm{d}(Q)}\big) = -\sqrt{\mathrm{d}(Q)}\,.$$

Hint. Choose homogeneous coordinates $(x_0 : x_1 : x_2 : x_3)$ on $\mathbb{P}^3$ so that Q is given by the equation

$$\sum_{i=0}^{3} a_i x_i^2 = 0\,.$$

Since the group of (projective) automorphisms of the quadric $Q_{K^{sep}}$ acts transitively on the set of lines on $Q_{K^{sep}}$, we may assume that the line l passes through the points $(\sqrt{a_1} : \sqrt{-a_0} : 0 : 0)$ and $(0 : 0 : \sqrt{a_3} : \sqrt{-a_2})$. Note that

$$\mathrm{d}(Q) = (-a_0)a_1(-a_2)a_3.$$

Therefore, the equality $g\big(\sqrt{\mathrm{d}(Q)}\big) = -\sqrt{\mathrm{d}(Q)}$ holds if and only if g changes the signs of an odd number of elements among $\sqrt{-a_0}$, $\sqrt{a_1}$, $\sqrt{-a_2}$, and $\sqrt{a_3}$. A straightforward check shows that this condition is equivalent to the fact that $g(l) \cap l$ is a point.

(ii) Use part (i) to solve Exercise 6.1.4(ii) without applying Exercise 6.1.4(i).

Hint. Suppose that the discriminant $\mathrm{d}(Q)$ is non-trivial and that $C(L) \neq \varnothing$. Then Exercise 6.1.2(v) implies that the quadric Q_L

contains some line l. Let σ be the non-trivial element of the Galois group $\mathrm{Gal}(L/K)$. Then

$$\sigma\left(\sqrt{\mathrm{d}(Q)}\right) = -\sqrt{\mathrm{d}(Q)}\,.$$

By part (i) this gives a K-point $l \cap \sigma(l)$ on Q.

Kummer theory (see Exercise 2.4.2(ii)) identifies the group $K^*/(K^*)^2$ with $\mathrm{Hom}(G_K, \mathbb{Z}/2\mathbb{Z})$. By Exercise 6.2.1(i), the homomorphism from G_K to $\mathbb{Z}/2\mathbb{Z}$ that corresponds to the discriminant $\mathrm{d}(Q)$ is defined by the images under the action of the Galois group G_K of two ample generators of the Picard group

$$\mathrm{Pic}(Q_{K^{sep}}) \cong \mathbb{Z} \oplus \mathbb{Z}\,.$$

A similar assertion holds for all even-dimensional quadrics after replacing the Picard group by the Chow group of algebraic cycles of middle dimension on the quadric.

Consider an orthogonal Grassmannian of the quadratic form q. In other words, consider the subvariety $\mathrm{G}(q)$ in the Grassmannian $\mathrm{G}(V,2)$ that parameterizes two-dimensional isotropic subspaces in V with respect to the form q, that is, the lines on Q. Over the separable closure K^{sep}, the variety $\mathrm{G}(q)_{K^{sep}}$ is isomorphic to a disjoint union of two copies of $\mathbb{P}^1$. Recall that choosing a morphism $B \to \mathrm{G}(V,2)$ of varieties over K is the same as choosing a subvariety $F \subset \mathbb{P}(V) \times B$ over K such that the corresponding projection to B is flat and its fibers are lines in $\mathbb{P}(V) \cong \mathbb{P}^3$. In particular, an identity morphism of the Grassmannian corresponds to the incidence variety $I \subset \mathbb{P}(V) \times \mathrm{G}(V,2)$.

EXERCISE 6.2.2. **Geometric meaning of the Clifford invariant**

(i) Show that the discriminant $\mathrm{d}(Q)$ is trivial if and only if the Grassmannian $\mathrm{G}(q)$ is reducible over K. In this case there is an isomorphism

$$\mathrm{G}(q) \cong C \sqcup C$$

of varieties over K, where the conic C is an arbitrary smooth plane section of the quadric Q over K.

Hint. Assume that the discriminant is trivial, and apply Exercise 6.1.2(v). Use the graphs of the morphisms π and π' in $Q \times C \subset \mathbb{P}^3 \times C$ to identify $C \sqcup C$ with the Grassmannian. For the converse implication, use an isomorphism

$$\mathrm{G}(q)_{K^{sep}} \cong \mathbb{P}^1 \sqcup \mathbb{P}^1$$

to show that

$$\mathrm{G}(q) \cong C' \sqcup C''\,,$$

where C' and C'' are smooth conics over K. Then consider the restriction $I_{C'}$ of the incidence variety I to $C' \subset \mathrm{G}(q)$. Since $I_{C'}$ projects isomorphically to $Q \subset \mathbb{P}(V)$, this gives a morphism $Q \to C'$. Similarly, the restriction of I to C'' gives a projection from Q to C''. The fibers of both projections are lines, which implies that the conics C' and C'' are isomorphic to each other and at the same time are isomorphic to an arbitrary smooth plane section C of the quadric Q; cf. Exercise 6.1.2. Therefore, we obtain an isomorphism $Q \cong C \times C$ (the fact that it is indeed an isomorphism can be checked over the algebraic closure of the field K). Now it remains to apply Exercise 6.1.2(viii).

(ii) Use part (i) to solve Exercise 6.2.1(i).

Hint. To prove that two homomorphisms from some group to $\mathbb{Z}/2\mathbb{Z}$ coincide, it is enough to check that they have the same kernel. This can be done using part (i) for the corresponding quadratic extensions of the field K.

(iii) Suppose that the discriminant $\mathrm{d}(Q)$ is non-trivial. Put $L = K\big(\sqrt{\mathrm{d}(Q)}\big)$ and let C be a conic over L that represents the Clifford invariant $\mathrm{cl}(Q_L) \in \mathrm{Br}(L)$. Any variety Y over L has the structure of a variety over K arising from the composition of morphisms

$$Y \to \mathrm{Spec}(L) \to \mathrm{Spec}(K)\,. \tag{6.1}$$

Consider this kind of structure of a variety over K on the conic C. Prove that there is an isomorphism $C \cong \mathrm{G}(q)$ of varieties over K.

Hint. By Exercise 6.1.2(v) the projection $\pi\colon Q_L \to C$ gives a morphism $j\colon C \to \mathrm{G}(q)_L$ of varieties over L. Since the restriction of scalars defined by the morphism (6.1) is left adjoint to the scalar extension, this gives a morphism $f\colon C \to G(q)$ of varieties over K. To show that f is an isomorphism, it is enough to prove that its scalar extension

$$f_L\colon C_L \to G(q)_L$$

is an isomorphism. Let σ be the non-trivial element in the Galois group $\mathrm{Gal}(L/K)$. Then

$$C_L \cong C \sqcup \sigma_* C\,,$$

where $\sigma_* C$ is defined as in Example 2.1.2(v). The map f_L corresponds to a pair of morphisms j and $\sigma_* j$, while the morphism

$$\sigma_* j\colon \sigma_* C \to \sigma_*\big(G(q)_L\big) \cong G(q)_L$$

is given by the projection

$$\sigma_*\pi\colon \sigma_*(Q_L) \cong Q_L \to \sigma_* C\,.$$

Since the fibers of $\sigma_*\pi$ are obtained by applying σ to the fibers of the initial projection π, we deduce from Exercise 6.2.1(i) that the fibers of π and $\sigma_*\pi$ are contained in different families. Therefore, f_L is an isomorphism.

6.3. Degenerations of quadrics

Let K be a complete discrete valuation field with discrete valuation v, valuation ring $\mathcal{O}_K$, and a perfect residue field κ of characteristic different from 2. Let $Q \subset \mathbb{P}^3$ be a smooth two-dimensional quadric over K.

EXERCISE 6.3.1. **Clifford invariant is unramified**

Suppose that the quadric Q has a projective model $\mathcal{Q} \subset \mathbb{P}^3$ over $\mathcal{O}_K$ that agrees with the initial projective embedding of Q (see §4.5) such that its closed fiber over the field κ is either a smooth quadric or a cone over a smooth conic (cf. the notion of a good model of a conic introduced before Exercise 4.5.4). Put

$$L = K\big(\sqrt{\mathrm{d}(Q)}\big)\,.$$

Prove that

$$\mathrm{res}\,\big(\,\mathrm{cl}(Q_L)\big) = 0\,.$$

Hint. Consider the class $\delta \in \mathrm{Br}(K)$ of an arbitrary smooth plane section of the quadric Q. Show that $\mathrm{cl}(Q_L)$ is the image of the element δ under the natural map

$$\mathrm{Br}(K) \to \mathrm{Br}(L)\,.$$

The main idea of the argument is that under our assumptions on degeneration of the model $\mathcal{Q}$ it has a smooth plane section with a smooth closed fiber. Specifically, since 2 is invertible in $\mathcal{O}_K$, the quadratic form q can be written in the diagonal form

$$q(x) = \sum_{i=0}^{3} a_i x_i^2, \quad a_i \in \mathcal{O}_K\,.$$

By the assumption on degeneration we may suppose that $v(a_i) = 0$ for $1 \leqslant i \leqslant 3$. Now let δ be the class of the conic given by the equations $q(x) = 0$ and $x_0 = 0$. Use Exercises 3.3.2(i) and 4.2.3(iv) to prove that $\mathrm{res}(\delta) = 0$, and recall that the diagram from Exercise 4.2.4(iv) commutes.

6.4. Further reading

One of the ways to study quadratic forms of arbitrary rank over a field (or, more generally, over a ring) is to consider the Witt ring generated by isomorphism classes of quadratic forms. Two-dimensional Pfister quadrics as in Exercise 6.1.2(i) and the corresponding Pfister forms of rank four have their natural higher-rank generalizations. One can find in [**MH73**] and [**Lam05**] excellent introductions to Witt rings, many useful facts about Pfister forms, a higher-rank generalization of the Clifford invariant, and a description of quadratic forms over local and global fields. We also recommend the book [**Kah08**].

Quadrics over non-closed fields were studied extensively from the point of view of algebraic geometry, in particular from the perspective of algebraic cycles and the theory of motives. See, for example, the collection of papers [**Izh04**] and the book [**EKM08**].

CHAPTER 7

Non-rational Double Covers of $\mathbb{P}^3$

In this chapter we will use unramified Brauer groups to prove non-rationality of certain unirational threefolds. These threefolds are double covers of the projective space $\mathbb{P}^3$ branched over (singular) quartics. The corresponding construction first appeared in the paper [**AM72**] of M. Artin and D. Mumford , but our exposition is based on a simpler approach due to M. Gross; see [**AM96**, Appendix].

As before, the obstruction to stable rationality of a variety X will come from (non-triviality of) the unramified Brauer group $\mathrm{Br}^{nr}\big(k(X)\big)$ of the field $k(X)$. In our case we will be able to simplify the computations (see Exercise 7.2.2) using the fact that the group $\mathrm{Br}^{nr}\big(k(X)\big)$ coincides with the unramified Brauer group $\mathrm{Br}^{nr}(X)$ of the variety X. The group $\mathrm{Br}^{nr}(X)$ is a bit more convenient than $\mathrm{Br}^{nr}\big(k(X)\big)$, because it allows us to consider only those valuations on the field $k(X)$ that correspond to divisors $D \subset X$.

7.1. More on the unramified Brauer group

Let X be a variety of dimension at least 2 over a field of characteristic different from 2. Recall that an isolated singular point x of the variety X is said to be an *ordinary double point* if it has an open neighborhood U in Zariski topology which is isomorphic to a germ of a hypersurface singularity with non-vanishing determinant of the Hessian matrix. In other words, there is an isomorphism

$$(U, x) \cong (\{f(t) = 0\}, 0) \subset \mathbb{A}^{n+1},$$

where $f(0) = 0$, the linear part of the polynomial f is zero, and the quadratic part of f is a non-degenerate quadratic form. Such a singularity can be resolved by a single blow-up $\widetilde{X} \to X$ with center at x; the blow-up replaces the point x in X by a smooth quadric D. By Serre's criterion for normality (see [**Har77**, Proposition II.8.23(b)]), a variety that has only ordinary double points is normal.

In what follows, we will need the following fact about unramified Brauer groups of varieties (see Definition 4.4.2). Let D be a smooth prime divisor in a smooth irreducible variety $\widetilde{X}$ over a field k of characteristic zero, so that we have a well-defined residue for elements of Brauer groups (see Exercise 4.2.1 and §4.3). Then for every element

$$\alpha \in \mathrm{Br}^{nr}\big(\widetilde{X} \setminus D\big) \subset \mathrm{Br}\big(k(\widetilde{X})\big),$$

its residue

$$\mathrm{res}_D(\alpha) \in \mathrm{Hom}(G_{k(D)}, \mathbb{Q}/\mathbb{Z})$$

corresponds to some unramified cyclic cover of D. One can show this using the interpretation of unramified Brauer groups in terms of étale cohomology of the sheaf of roots of unity (see §A.5 and §A.6), together with the exact sequence of localization for étale cohomology with finite coefficients; see [**Mil80**, VI.5.4(b)].

EXERCISE 7.1.1. **Unramified Brauer group and ordinary double points**
Let X be an irreducible projective variety over a field k of characteristic zero such that all singular points of X are ordinary double points. Prove that the natural embedding

$$\mathrm{Br}^{nr}\big(k(X)\big) \hookrightarrow \mathrm{Br}^{nr}(X)$$

is an isomorphism.

Hint. Consider the resolution of singularities $\widetilde{X}$ of the variety X obtained by blowing up the singular points, and use a similar fact that we know to hold for smooth varieties together with the observation that a smooth quadric does not have non-trivial unramified covers.

7.2. Families of two-dimensional quadrics

Let B be a normal irreducible variety over a field k of characteristic different from 2. Denote by $K = k(B)$ the field of rational functions on B. Let $\mathcal{Q}$ be a family of quadrics in $\mathbb{P}^3$ with base B such that its fiber at the generic schematic point of B is a smooth quadric Q over K. Therefore,

$$\mathcal{Q} \subset \mathbb{P}^3_B = \mathbb{P}^3 \times B$$

is a projective model of the quadric Q that agrees with its projective embedding over K (see §4.5). One can show that $\mathcal{Q}$ can be realized as a subvariety in $\mathbb{P}^3 \times B$ given by zeros of a section of a line bundle

$$\mathcal{O}_{\mathbb{P}^3}(2) \boxtimes \mathcal{L},$$

where $\mathcal{L} \in \mathrm{Pic}(B)$ and the morphism $\mathcal{Q} \to B$ is flat with a smooth generic fiber. Here $\boxtimes$ denotes a tensor product on $\mathbb{P}^3 \times B$ of pull-backs of two line bundles with respect to the projections $\mathbb{P}^3 \times B \to \mathbb{P}^3$ and $\mathbb{P}^3 \times B \to B$.

Let $\Delta^i \subset B$ be a subvariety that consists of points such that the corresponding fiber of $\mathcal{Q}$ is a quadric of corank at least i. In particular, Δ^1 is a divisor on B, because Δ^1 is given locally by a single equation (what is this equation?). Suppose that Δ^1 is non-empty (for a projective B this is equivalent to the assumption that the family of quadrics $\mathcal{Q}$ is non-trivial) and that $\mathrm{codim}_B(\Delta^2) \geqslant 2$. Suppose also that the discriminant $\mathrm{d}(Q)$ (see Definition 6.1.1) is non-trivial.

Put

$$L = K\big(\sqrt{\mathrm{d}(Q)}\big)$$

and denote by X the normalization of B in L. Thus X is a normal irreducible variety over the field k with a field of rational functions $k(X) \cong L$, and there is a surjective morphism $X \to B$ of degree 2 corresponding to the field extension $K \subset L$ over the generic point.

EXERCISE 7.2.1. **Singularities of the double cover**

(i) Prove that Δ^1 is the branch divisor of the finite surjective morphism $X \to B$.
(ii) Suppose that the variety B is smooth and all singularities of the divisor Δ^1 are ordinary double points. Show that all singularities of X are ordinary double points as well.

Hint. Look at the equation that defines X locally as a divisor in $B \times \mathbb{A}^1$.

EXERCISE 7.2.2. **Non-triviality of the unramified Brauer group**
Suppose that B is a smooth projective variety, singularities of the divisor Δ^1 are ordinary double points, and $Q(K) = \varnothing$. Put (see Definition 6.1.3)

$$\alpha = \mathrm{cl}(Q_L) \in \mathrm{Br}(L)\,.$$

(i) Show that $\alpha \neq 0$.

Hint. Apply Exercise 6.1.4(iii).

(ii) Until the end of the exercise assume that k is of characteristic zero, so that we can speak about residues of elements of Brauer groups (see Exercise 4.2.1 and §4.3). Show that for every prime divisor D on X one has $\mathrm{res}_D(\alpha) = 0$, that is,

$$\alpha \in \mathrm{Br}^{nr}(X)\,.$$

Hint. Consider the valuation on the field K given by the image of the divisor D in B, and use the lower bound

$$\mathrm{codim}_B(\Delta^2) \geqslant 2$$

for the codimension of Δ^2 in B together with Exercise 6.3.1.

(iii) Show that

$$\alpha \in \mathrm{Br}^{nr}(L)\,.$$

Hint. Use part (ii) and Exercises 7.2.1(ii) and 7.1.1.

(iv) Show that the variety X is not stably rational.

7.3. Construction of a geometric example

Let V be a vector space of dimension $n \geqslant 3$ over a field k of characteristic different from 2. For every $i \geqslant 0$ define

$$\Sigma^i \subset \mathbb{P}\big(\mathrm{Sym}^2(V^\vee)\big)$$

as the subset of points that correspond to quadratic forms of corank at least i on V. Here $V^\vee$ denotes the dual space to V and Sym^2 stands for the symmetric square of a vector space; thus $\mathrm{Sym}^2(V^\vee)$ is the space of all symmetric bilinear forms on V. In particular, Σ^1 is a divisor given by the equation $\det(q) = 0$, where $q \in \mathrm{Sym}^2(V^\vee)$.

EXERCISE 7.3.1. **Singularities of the variety of degenerate quadrics**

(o) Consider the natural action of the group $\mathrm{GL}(V)$ on $\mathbb{P}\big(\mathrm{Sym}^2(V^\vee)\big)$. The subvarieties Σ^i are invariant with respect to this action. Check that for every $i \geqslant 0$ the set $\Sigma^i \setminus \Sigma^{i+1}$ is a single $\mathrm{GL}(V)$-orbit, which is dense in Σ^i.

Hint. Diagonalize a quadratic form.

(i) Check that Σ^1 is a prime divisor of degree n.

Hint. The subvariety Σ^1 is given by a single equation $\det(-) = 0$. Irreducibility of Σ^1 follows from part (o) and irreducibility of the variety $\mathrm{GL}(V)$. Note also that the equation $\det(-) = 0$ defines a reduced divisor; this can be checked using its differential at a convenient point $q \in \Sigma^1$.

(ii) Prove that
$$\operatorname{codim}_{\Sigma^1}(\Sigma^2) = 2$$
and Σ^2 is the singular locus of Σ^1.

Hint. Put
$$I_2 = \operatorname{diag}(0, 0, \underbrace{1, \ldots, 1}_{n-2}).$$
By part (o) it is enough to compute the dimension of Σ^2 at the point I_2 and check that Σ^1 is singular at this point. The former can be done using an explicit description of the stabilizer of the matrix I_2 in $\mathrm{GL}(V)$. The latter follows from an explicit computation of partial derivatives of the determinant.

(iii) Let $\mathfrak{H}_q$ be the matrix of second partial derivatives of the function $\det(-)$ at a point
$$q \in \mathbb{P}\big(\operatorname{Sym}^2(V^\vee)\big).$$
Prove that $\mathfrak{H}_q$ has rank 3 provided that q is a general point of Σ^2.

Hint. It is enough to check this at the point I_2 of Σ^2.

Let V be a four-dimensional vector space over the field $\mathbb{C}$ of complex numbers. Let B be a general three-dimensional subspace in the projective space
$$\mathbb{P}\big(\operatorname{Sym}^2(V^\vee)\big) \cong \mathbb{P}^9.$$
Let the five-dimensional variety $\mathcal{Q}$ be the corresponding fibration in quadrics in $\mathbb{P}(V)$ with base B, and let Q be the fiber of $\mathcal{Q}$ at the generic schematic point of B. In particular, Q is a smooth quadric over the field $K = \mathbb{C}(B)$. For an algebraic variety U over the field $\mathbb{C}$, we denote by $H^i\big(U(\mathbb{C}), \mathbb{Z}\big)$ the ith Betti cohomology group of the corresponding topological space $U(\mathbb{C})$ with classical topology. Recall that there is a product
$$H^i\big(U(\mathbb{C}), \mathbb{Z}\big) \times H^j\big(U(\mathbb{C}), \mathbb{Z}\big) \to H^{i+j}\big(U(\mathbb{C}), \mathbb{Z}\big).$$
If U is irreducible and smooth, then every (irreducible) subvariety $Z \subset U$ of codimension i has a well-defined class $[Z] \in H^{2i}\big(U(\mathbb{C}), \mathbb{Z}\big)$. Finally, if U is also projective, there is a canonical isomorphism $H^{2d}\big(U(\mathbb{C}), \mathbb{Z}\big) \cong \mathbb{Z}$, where d is the dimension of U.

EXERCISE 7.3.2. **Stable non-rationality of a special double cover of $\mathbb{P}^3$ branched over a quartic**

(o) Show that equation $\det(Q) = 0$ defines a prime divisor Δ^1 of degree 4 in B. Check that Δ^2 is a finite collection of points, Δ^2 is the singular locus of the divisor Δ^1, and the singularities of Δ^1 are ordinary double points. Prove that the discriminant $\mathrm{d}(Q)$ is non-trivial.

Hint. All these assertions except for the last one follow from Exercise 7.3.1 and Bertini's theorem. To prove that the discriminant is non-trivial, compute the valuation corresponding to the divisor Δ^1 on $\mathrm{d}(Q)$.

(i) Prove that the variety $\mathcal{Q}$ is a smooth divisor of bidegree $(1, 2)$ in a smooth projective variety
$$W = B \times \mathbb{P}(V).$$

Hint. Use the following form of Bertini's theorem (see e.g. [**Har95**, Theorem 17.16]): if $\psi\colon M \to \mathbb{P}^n$ is an arbitrary morphism from a smooth variety M, then the preimage of a general hyperplane with respect to ψ is smooth. This theorem should be applied to the projection of the incidence variety

$$I \subset \mathbb{P}\big(\mathrm{Sym}^2(V^\vee)\big) \times \mathbb{P}(V)$$

to the first factor.

(ii) Let $\mathcal{Q}_b$ be a fiber of $\mathcal{Q}$ at an arbitrary closed point $b \in B$. Prove that for every $c \in H^4\big(\mathcal{Q}(\mathbb{C}), \mathbb{Z}\big)$ the product

$$c \cdot [\mathcal{Q}_b] \in H^{10}\big(\mathcal{Q}(\mathbb{C}), \mathbb{Z}\big) \cong \mathbb{Z}$$

is even.

Hint. First prove a similar assertion for an arbitrary element

$$c' \in H^4\big(W(\mathbb{C}), \mathbb{Z}\big)$$

and the class $\mathcal{Q}_b$ in $H^8\big(W(\mathbb{C}), \mathbb{Z}\big)$ using part (i). Then apply the Lefschetz hyperplane section theorem to find an element $c' \in H^4\big(W(\mathbb{C}), \mathbb{Z}\big)$ whose restriction to $\mathcal{Q}$ equals c.

(iii) Prove that $Q(K) = \varnothing$.

Hint. Assuming that there is a point in $Q(K)$, consider the Zariski closure of the graph of the corresponding rational map $B \dashrightarrow \mathcal{Q}$ and use part (ii).

(iv) Suppose that the double cover $X \to B$ is constructed as in §7.2. Show that the variety X is not stably rational.

Hint. Use Exercise 7.2.2(iv).

(v) Replace $B \cong \mathbb{P}^3$ by $B \cong \mathbb{P}^2$ in the above discussion. What changes?

7.4. Some unirationality constructions

Compared to other parts of the book, this section requires some additional knowledge of algebraic geometry: in particular, we will assume that the reader is familiar with such notions as an embedding given by a linear system and intersection theory on surfaces. All the necessary background can be found, for instance, in [**Har77**]. Note that we are not going to use the results of this section in the rest of the book.

Let F be a field of characteristic different from 2 and $\bar{F}$ its algebraic closure. Let S be a smooth del Pezzo surface of degree 2 over F, that is, a smooth surface with an ample anticanonical divisor $-K_S$ such that $K_S \cdot K_S = 2$. Then the surface $S_{\bar{F}}$ is isomorphic to a blow-up of seven $\bar{F}$-points on $\mathbb{P}^2$; in particular, $S_{\bar{F}}$ is rational. The anticanonical linear system defines a morphism

$$\varphi = \varphi_{|-K_S|}\colon S \to \mathbb{P}^2,$$

which is a double cover of $\mathbb{P}^2$ branched over a smooth quartic curve $T \subset \mathbb{P}^2$. Recall that *lines* on a del Pezzo surface of degree 2 (or, more generally, on a del Pezzo surface of degree at most 2) are the curves of anticanonical degree 1. The surface $S_{\bar{F}}$ contains a finite number of lines and they are smooth curves of genus zero (which is exactly what one would expect from a line). Note that on a del Pezzo surface of degree 1, a general anticanonical curve is an elliptic curve of anticanonical degree 1;

this is why we have to impose the lower bound on the degree of a del Pezzo surface to use our definition of a line. More background on del Pezzo surfaces is available in [**Man86**, Chapter IV], [**Dol12**, Chapter 8], and [**Kol96**, § III.3].

The following exercise mostly follows the argument from the proof of [**Man86**, Theorem IV.7.7]. A more general and more detailed treatment can be found in [**STVA14**]. Let R be the ramification divisor of the morphism φ, that is, the preimage on S of the branch curve $T \subset \mathbb{P}^2$ of φ.

EXERCISE 7.4.1. **Unirationality of a del Pezzo surface of degree 2 with a point**

Let $U \subset S$ be the complement of the union of the curve R and all lines on $S_{\bar{F}}$. Note that U is a Zariski-open subset of S.

(o) Let $l \subset \mathbb{P}^2$ be a line (defined over $\bar{F}$). Check that the preimage

$$\tilde{l} = \varphi^{-1}(l) \subset S_{\bar{F}}$$

is reducible if and only if $\tilde{l}$ splits into a union of two lines on $S_{\bar{F}}$.

(i) Prove that for every point $x \in S(F)$ not contained in any line on $S_{\bar{F}}$ there is a curve

$$C \in |-2K_S - 3x|,$$

that is, a curve from the linear system $|-2K_S|$ that has a singular point of multiplicity at least 3 at x.

Hint. It is enough to work over the algebraic closure of the ground field in order to check that a linear system is not empty. Thus we can assume from the very beginning that the field F is algebraically closed. We have a birational morphism $\pi\colon S \to \mathbb{P}^2$ that is a blow-up of seven points $P_1, \ldots, P_7$ on $\mathbb{P}^2$ with exceptional divisors $E_1, \ldots, E_7$. The point x is not contained in any of the exceptional divisors E_i. Recall that one has a linear equivalence

$$K_S \sim \pi^* K_{\mathbb{P}^2} + \sum_{i=1}^{7} E_i \,.$$

Moreover, a divisor D on $\mathbb{P}^2$ has multiplicity at least 2 at the point P_i if and only if the divisor $\pi^* D$ has multiplicity at least 2 along E_i. Finally, since the point x is not contained in any of the divisors E_i, the morphism π is an isomorphism in a neighborhood of x. Hence there is a map

$$|-2K_{\mathbb{P}^2} - 2\sum_{i=1}^{7} P_i - 3\pi(x)| \longrightarrow |-2K_S - 3x|$$

given by the formula

$$D \mapsto \pi^{-1}(D) + \sum_{i=1}^{7} \bigl(\operatorname{mult}_{P_i}(D) - 2\bigr) E_i \,.$$

One can check that this map is bijective, but we will not use this fact. The class $-2K_{\mathbb{P}^2}$ is the linear equivalence class of curves of degree 6 on $\mathbb{P}^2$. The requirement that a curve have multiplicity at least 2 (respectively, at least 3) at a given point imposes 3 (respectively, 6) linear conditions on the equation of the curve. Thus the dimension of the linear system

$$|-2K_{\mathbb{P}^2} - 2\sum_{i=1}^{7} P_i - 3\pi(x)|$$

is at least

$$27 - 7 \cdot 3 - 6 = 0\,.$$

Therefore, this linear system is not empty, and hence the linear system $|-2K_S - 3x|$ is not empty either.

Actually, the above assertion also holds if the point x is contained in one of the lines on S, but proving it requires more effort in this case.

(ii) Show that if $x \in S(F)$ is not contained in any line on $S_{\bar{F}}$, then no curve passing through x can contain an irreducible component (defined over $\bar{F}$) that is isomorphically mapped to a line passing through the point $\varphi(x)$ by the anticanonical morphism $\varphi\colon S \to \mathbb{P}^2$.

Hint. Use part (o).

(iii) Show that if x is contained in $U(F)$, then every curve C from the linear system $|-2K_S - 3x|$ is irreducible (even over $\bar{F}$) and F-rational, and the multiplicity $\mathrm{mult}_x(C)$ of C at the point x equals 3.

Hint. The anticanonical image $\varphi(C) \subset \mathbb{P}^2$ of the curve C is a curve of degree at most 4. Since $x \notin R$, in a neighborhood of x the morphism φ is unramified and thus

$$\mathrm{mult}_{\varphi(x)}\big(\varphi(C)\big) \geqslant \mathrm{mult}_x(C) \geqslant 3\,.$$

If $\mathrm{mult}_{\varphi(x)}\big(\varphi(C)\big) = 4$ or $\deg\big(\varphi(C)\big) = 3$, then $\varphi(C)$ is a union of, respectively, four or three lines over $\bar{F}$ passing through $\varphi(x)$. This is impossible by part (ii). So we see that $\mathrm{mult}_{\varphi(x)}\big(\varphi(C)\big) = 3$ and $\deg\big(\varphi(C)\big) = 4$. Hence $\mathrm{mult}_x(C) = 3$ and C is mapped birationally on $\varphi(C)$. Suppose that C is reducible. Then $\varphi(C)$ is also reducible and the above information about multiplicity and degree implies that $\varphi(C)$ contains a line over $\bar{F}$ passing through $\varphi(x)$. This again gives a contradiction to part (ii). Therefore, the curve C is irreducible and birational to $\varphi(C)$. Since $\mathrm{mult}_{\varphi(x)}\big(\varphi(C)\big) = 3$ and $\deg\big(\varphi(C)\big) = 4$, we conclude that the curve $\varphi(C)$ is rational, which means that the curve C is rational as well.

(iv) Show that if $x \in U(F)$, then a curve $C \in |-2K_S - 3x|$ is unique.

Hint. Suppose that there are two different curves

$$C_1, C_2 \in |-2K_S - 3x|\,.$$

Then C_1 and C_2 are irreducible by part (iii), so that

$$8 = C_1 \cdot C_2 \geqslant \mathrm{mult}_x(C_1) \cdot \mathrm{mult}_x(C_2) = 9\,,$$

a contradiction.

We will denote this curve by C_x. If the point x is defined not over the field F but over some field $F' \supset F$, then the curve C_x is also defined over the field F'.

(v) Suppose that the surface S has an F-point

$$x \in U(F) \subset S(F).$$

Construct the curve C_x as described in part (iv). Furthermore, for a general point $y \in C_x$ construct the curve C_y in a similar way. Consider the incidence variety

$$\widetilde{S} = \{(y, z) \mid y \in C_x \cap U, z \in C_y\}.$$

Prove that the surface $\widetilde{S}$ is rational over the field F.

Hint. The projection of the surface $\widetilde{S}$ to $C_x \cap U$ gives a fibration structure on $\widetilde{S}$. The generic fiber of this fibration is isomorphic to a curve C_η over the field $F(C_x)$ that is constructed from the generic point η of the curve C_x as described in part (iv). It follows from part (iii) that the curve C_η is rational over the field $F(C_x)$, that is, over the field of rational functions in one variable over F.

(vi) Suppose that the surface S has an F-point not contained in any line on $S_{\bar{F}}$ and not contained in the ramification curve $R \subset S$. Prove that S is unirational over F.

Hint. Show that the morphism

$$\widetilde{S} \to S, \quad (y, z) \mapsto z$$

is dominant. This follows from the fact that the surface $\widetilde{S}$ is irreducible and its image contains at least two irreducible curves, namely, C_x and C_y for a general enough point $y \in C_x$.

(vii) Prove F-unirationality of a del Pezzo surface of the degree $d \geqslant 2$ that is defined over F and has enough (say, a Zariski-dense set of) F-points.

Hint. Blow up some points!

The reader with more advanced knowledge of the geometry of del Pezzo surfaces can construct the curve C from Exercise 7.4.1(i) using the following geometric argument. Let $\pi\colon S' \to S$ be the blow-up of the point x with an exceptional divisor E. Since x is not contained in any line on $S_{\bar{F}}$, one can check that S' is a del Pezzo surface of degree 1. The linear system $|-2K_{S'}|$ gives a morphism

$$\psi\colon S' \to Q$$

which is a double cover of a quadric cone Q (see e.g. [**Dol12**, §8.8.2]). Consider the Galois involution

$$\iota\colon S' \to S'$$

of this double cover. The required curve C can be described as $\pi\big(\iota(E)\big)$. Irreducibility and rationality of C are obvious from this construction. Note also that the generality condition one has to impose on a point $x \in S$ can be significantly weakened; see [**STVA14**].

Now we will apply Exercise 7.4.1 to obtain a unirationality construction for the variety from Exercise 7.3.2.

EXERCISE 7.4.2. **Unirationality of a double cover of $\mathbb{P}^3$ branched over a quartic**

Let k be an algebraically closed field of characteristic different from 2, and let $V \subset \mathbb{P}^3$ be a surface of degree 4 defined over k. Suppose that V has isolated singularities (so that in particular it is irreducible). Let $\pi\colon X \to \mathbb{P}^3$ be the double cover of $\mathbb{P}^3$ branched over the surface V; this is a normal irreducible variety.

(o) Prove that X contains a two-parameter family of lines, that is, curves that are isomorphically mapped to lines in $\mathbb{P}^3$ by the map π.

Hint. The image of a line on X under the map π is a bitangent line of the surface $V \subset \mathbb{P}^3$. The dimension of the family of bitangent lines to V can be computed using the natural embedding of this family as a subvariety of the Grassmannian $\mathrm{G}(4,2)$ of two-dimensional subspaces in a four-dimensional vector space.

(i) Consider the projection $\phi\colon \mathbb{P}^3 \dashrightarrow \mathbb{P}^1$ from a general line L in $\mathbb{P}^3$. It gives a rational map $\theta\colon X \dashrightarrow \mathbb{P}^1$. Let $\psi\colon \widetilde{X} \to X$ be the blow-up of the preimage of the line L on X. Then ψ gives a resolution of indeterminacies of the rational map θ. The composition $\theta \circ \psi$ is a fibration $\widetilde{X} \to \mathbb{P}^1$. Let S be a surface over the field

$$F = k(\mathbb{P}^1) \cong k(t)$$

that is the generic schematic fiber of the morphism $\theta \circ \psi$. Show that S is a smooth del Pezzo surface of degree 2.

(ii) Show that the set of F-points is Zariski dense in the surface S.

Hint. Use part (o) to construct many sections of the fibration $\theta \circ \psi\colon \widetilde{X} \to \mathbb{P}^1$.

Combining this with Exercise 7.4.1, deduce that the surface S is unirational.

(iii) Prove that the variety $\widetilde{X}$ (and thus also the variety X) is unirational.

Hint. Use part (ii).

7.5. Further reading

The torsion subgroup in the third Betti cohomology with integral coefficients as an obstruction to stable rationality of smooth complex projective varieties was known since the work of [**AM72**], M. Artin and D. Mumford. Nevertheless, for a long time only a handful of examples of (interesting) varieties with a non-trivial invariant of this kind was known. Some of them were constructed in the same way as in [**AM72**], or rather as in [**AM96**, Appendix], which provided a more explicit interpretation in terms of Brauer groups (cf. the exact sequence (A.8) in §A.7); see for instance [**IKP14**], [**Huh13**], and [**PS16**, §5]. A few exceptions include the construction of [**DG94**, §5] using the 3-torsion instead of 2-torsion (unlike the papers mentioned above). See also [**Zag77**] for more details on the torsion subgroup in the cohomology.

However, recent years have seen significant progress in studying obstructions for stable rationality. The groundbreaking work of C. Voisin [**Voi15**] that connected the Artin–Mumford approach to universal triviality of the Chow group CH_0 initiated much active research in this direction. Fano varieties treated by this method include quartic threefolds (see [**CTP16b**]); certain cyclic covers ([**CTP16a**], [**Oka16a**]); higher-dimensional hypersurfaces of relatively large degree, including a four-dimensional quartic (see [**Tot16**]); non-rational smooth Fano threefolds, with an exception of cubic hypersurfaces (see [**HT16**]); and other types of varieties (e.g. [**Bea16**], [**HPT16b**], [**Oka16b**], [**Oka17a**], [**Oka17b**]). In general, this approach can be applied to prove stable non-rationality for a *very general* variety in the corresponding deformation family, that is, a variety from a complement of a countable union of Zariski-closed subsets in the moduli; however, in certain cases the approach can be adjusted to prove stable non-rationality of particular varieties; see [**Tot16**, §3].

Furthermore, there are various results concerning conic and quadric bundles (e.g. [**HKT16**], [**BvB16**], [**ABvBP16**], [**AO16**], [**HPT16a**], [**Sch17a**], [**Sch17b**]), and del Pezzo fibrations (e.g. [**KO17**], [**KT17b**]). Some results on stable non-rationality were used to study deformations of stably rational varieties; see [**Tot16**, §4], [**HPT16a**], [**HPT17**], and [**ABP17**]. See [**NS17**] for some general theory (cf. also [**KT17a**]).

We refer the reader to a nice survey [**Pir16**] for more details on the subject.

CHAPTER 8

Weil Restriction and Algebraic Tori

8.1. Weil restriction

Let $k \subset K$ be a finite separable extension of fields of degree d, and let X be a quasi-projective variety over K. Fix an embedding of the field K into k^{sep} over k. Let L be the normal closure of the field K in k^{sep} over k. There are d embeddings

$$\sigma_1, \ldots, \sigma_d \colon K \hookrightarrow L$$

over k. Applying the embeddings σ_i, we construct out of X a collection of varieties X_{σ_i} over the field L as follows: if X is given by equations, then, by applying σ_i to their coefficients, we obtain equations of X_{σ_i}. Note that the varieties X_{σ_i} are not necessarily isomorphic to each other, even over the separable closure $L^{sep} = k^{sep}$.

Consider the variety $X_{\sigma_1} \times \ldots \times X_{\sigma_d}$ over L. For any element $g \in \mathrm{Gal}(L/k)$, there is an isomorphism

$$g_*\big(X_{\sigma_1} \times \ldots \times X_{\sigma_d}\big) \cong X_{g\sigma_1} \times \ldots \times X_{g\sigma_d}$$

of varieties over L. Furthermore, upon permuting factors, we obtain an isomorphism

$$g_*\big(X_{\sigma_1} \times \ldots \times X_{\sigma_d}\big) \cong X_{\sigma_1} \times \ldots \times X_{\sigma_d}\,.$$

Thus we obtain canonical descent data on the variety $X_{\sigma_1} \times \ldots \times X_{\sigma_d}$ with respect to the field extension $k \subset L$. According to Exercise 2.1.4(iv), this gives a uniquely defined quasi-projective variety Y over k such that its scalar extension Y_L is isomorphic to $X_{\sigma_1} \times \ldots \times X_{\sigma_d}$.

DEFINITION 8.1.1. The variety Y is called the *Weil restriction* of the variety X with respect to the field extension $k \subset K$ and is denoted by $R_{K/k}(X)$.

Weil restriction is an algebraic analog of the analytic procedure that allows one to consider an n-dimensional complex manifold as a $2n$-dimensional real manifold; cf. Exercises 8.1.2(iv), 8.1.3(i), and 8.1.6(iii).

EXERCISE 8.1.2. **Functorial properties of Weil restriction**

(o) Check that Weil restriction defines a functor $R_{K/k}$ from the category of quasi-projective varieties over the field K to the category of quasi-projective varieties over the field k.

(i) Let X_1 and X_2 be two varieties defined over K. Prove that

$$R_{K/k}(X_1 \times_K X_2) \cong R_{K/k}(X_1) \times_k R_{K/k}(X_2)\,.$$

(ii) Prove that the Weil restriction of an algebraic group over K is an algebraic group over k.

Hint. Use part (i).

(iii) Prove that the Weil restriction functor is right adjoint to the scalar extension functor, that is, for any K-variety X and any k-variety Y, there is a canonical bijection

$$\mathrm{Hom}_K\left(Y_K, X\right) \cong \mathrm{Hom}_k\left(Y, R_{K/k}(X)\right),$$

where $\mathrm{Hom}_F(U, V)$ denotes the set of morphisms between algebraic varieties U and V over a field F.

(iv) Prove that there is a canonical bijection $X(K) \cong R_{K/k}(X)(k)$.

(v) Let Y be a variety over k. The action of the Galois group $\mathrm{Gal}(L/k)$ on the set of embeddings of K into L over k gives a homomorphism

$$\varphi\colon \mathrm{Gal}(L/K) \to \mathrm{S}_d \subset \mathrm{Aut}\left(Y^{\times d}\right).$$

Prove that the variety $R_{K/k}(Y_K)$ is the twist of the variety $Y^{\times d}$ by φ; see Exercise 2.2.1(i) and (iii).

Recall (see §3.3) that a variety X over a field k is said to be geometrically irreducible if the variety $X_{\bar{k}}$ is irreducible (and otherwise X is called geometrically reducible).

Exercise 8.1.3. **Geometric properties of Weil restriction**

(i) Prove that

$$\dim_k\left(R_{K/k}(X)\right) = [K : k] \cdot \dim_K(X).$$

(ii) Let $Z \subset X$ be a closed subvariety in X defined over K, and let $U = X \setminus Z$ be the open complement. Prove that $R_{K/k}(Z)$ is a closed subvariety in $R_{K/k}(X)$ and that $R_{K/k}(U)$ is an open subset in $R_{K/k}(X)$. Is it true that

$$R_{K/k}(X) \cong R_{K/k}(Z) \cup R_{K/k}(U)\,?$$

Hint. Consider the example $X = \mathbb{A}^1_K$ and $Z = \{0\}$, and extend scalars from k to k^{sep}.

(iii) Prove that the Weil restriction of a geometrically irreducible variety is also geometrically irreducible.

(iv) Is it true that the Weil restriction of an irreducible variety is irreducible?

Hint. Apply Exercise 8.1.2(v) to $Y = \mathrm{Spec}(E)$, where E is a finite Galois extension of k such that $K \otimes_k E$ does not have zero divisors.

(v) Show that there is a canonical bijection

$$R_{K/k}(X)(k^{sep}) \cong \mathrm{Map}_{G_K}\left(G_k, X(k^{sep})\right)$$

that commutes with the action of the group G_k, where G_k acts on the right-hand side by the formula

$$\left({}^g\varphi\right)(g') = \varphi(g'g), \quad \varphi \in \mathrm{Map}_{G_K}\left(G_k, X(k^{sep})\right), \quad g, g' \in G_k,$$

and G_K acts on G_k by left translations.

Hint. By definition, there is a canonical G_k-equivariant bijection

$$R_{K/k}(X)(k^{sep}) \cong X_{\sigma_1}(k^{sep}) \times \ldots \times X_{\sigma_d}(k^{sep}).$$

Suppose that σ_1 corresponds to the fixed embedding of K into k^{sep}. Then an element

$$(x_{\sigma_1}, \ldots, x_{\sigma_d}) \in X_{\sigma_1}(k^{sep}) \times \ldots \times X_{\sigma_d}(k^{sep})$$

corresponds to the map

$$g \mapsto g\big(x_{g^{-1}\sigma_1}\big), \quad g \in G_k\,,$$

and a map

$$\varphi \in \mathrm{Map}_{G_K}\big(G_k, X(k^{sep})\big)$$

corresponds to the collection of points

$$\big(g_1\varphi(g_1^{-1}), \ldots, g_d\varphi(g_d^{-1})\big)\,,$$

where the elements $g_i \in G_k$ are such that $g_i\sigma_1 = \sigma_i$.

(vi) Suppose that X is a commutative algebraic group over K. Show that there is an isomorphism of G_k-modules

$$R_{K/k}(X)(k^{sep}) \cong i_*X(k^{sep})\,,$$

where $i\colon G_K \hookrightarrow G_k$ denotes the natural embedding and i_* is the coinduction; see Definition 1.2.3.

Hint. Use part (v) and the remark after Definition 1.2.3.

EXERCISE 8.1.4. **Weil restriction of affine varieties**

(i) Prove that there is a (non-canonical) isomorphism

$$R_{K/k}(\mathbb{A}^1) \cong \mathbb{A}^d\,.$$

Hint. By Exercise 8.1.2(iii), for any k-variety Y there is a canonical bijection between the set $\mathrm{Hom}_k\big(Y, R_{K/k}(\mathbb{A}^1)\big)$ and the ring of regular functions

$$K[Y_K] \cong K \otimes_k k[Y]\,.$$

Furthermore, a choice of a basis in K over k defines a bijection of $K \otimes_k k[Y]$ with $k[Y]^{\oplus d}$, that is, with the set $\mathrm{Hom}_k(Y, \mathbb{A}^d)$.

(ii) Prove that $R_{K/k}(\mathbb{A}^n) \cong \mathbb{A}^{nd}$.

Hint. Use part (i) and Exercise 8.1.2(i).

(iii) Show that if a variety $X \subset \mathbb{A}^n$ is affine, then $R_{K/k}(X)$ is affine as well and is embedded into $\mathbb{A}^{nd}$.

Hint. Use part (ii) and Exercise 8.1.3(ii).

(iv) Let $X \subset \mathbb{A}^n$ be an affine hypersurface given by an equation

$$f(x_1, \ldots, x_n) = 0\,.$$

Prove that the affine variety $R_{K/k}(X) \subset \mathbb{A}^{nd}$ is given by the following d equations. Choose a basis $e_1, \ldots, e_d$ in K over k. Consider formal variables x_{ij}, where $1 \leqslant i \leqslant n$ and $1 \leqslant j \leqslant d$. Replace each x_i in f by the expression $\sum_{j=1}^d x_{ij}e_j$ using the relations

$$e_\alpha \cdot e_\beta = \sum_{\gamma=1}^{d} c_{\alpha\beta}^{\gamma} e_\gamma$$

that hold in the field K, where $c^{\gamma}_{\alpha\beta} \in k$. This gives an equality

$$f\left(\sum_{j=1}^{d} x_{1j}e_j, \ldots, \sum_{j=1}^{d} x_{nj}e_j\right) = \sum_{l=1}^{d} f_l(x_{ij})e_l$$

that defines d equations $f_l(x_{ij}) = 0$, $1 \leqslant l \leqslant d$.

Hint. Argue as in part (i).

(v) Let $X \subset \mathbb{A}^n$ be an affine variety. Check that the equations that define $R_{K/k}(X)$ are obtained from the equations that define X by the method described in part (iv).

EXERCISE 8.1.5. **Weil restriction of rational varieties**
Let X be a K-rational variety. Prove that the variety $R_{K/k}(X)$ is k-rational.

Hint. Use Exercises 8.1.4(ii) and 8.1.3(ii) and (iii).

EXERCISE 8.1.6. **Weil restriction of projective varieties**

(i) Let $X \subset \mathbb{P}^n = \mathbb{P}(W)$ be a projective variety over K. Construct a closed embedding

$$R_{K/k}(X) \subset \mathbb{P}^{(n+1)^d-1}$$

over k.

Hint. Each embedding $\sigma\colon K \hookrightarrow L$ over k defines an L-vector space $W \otimes_{K,\sigma} L$ obtained from W by taking scalar extension with respect to σ (that is, by taking scalar extension from $\sigma(K)$ to L). Consider the L-vector space

$$U = (W \otimes_{K,\sigma_1} L) \otimes_L \ldots \otimes_L (W \otimes_{K,\sigma_d} L)$$

and construct a closed embedding

$$X_{\sigma_1} \times \ldots \times X_{\sigma_d} \hookrightarrow \mathbb{P}(U)\,.$$

Then argue as in Exercise 2.1.5(iii).

(ii) Suppose that the characteristic of the field k is different from 2. Show that for a quadratic extension

$$k \subset K = k(\sqrt{a}), \quad a \in k^*\,,$$

the variety $R_{K/k}(\mathbb{P}^n)$ is given in the projectivization of the space of $(n+1) \times (n+1)$ matrices by the condition

$$\mathrm{rk}\left((A + A^T) + \sqrt{a}(A - A^T)\right) \leqslant 1\,,$$

where rk denotes the rank of a matrix and A^T is the transpose of a matrix A.

Hint. Let g be the non-trivial element of the group $G = \mathrm{Gal}(K/k)$. For any k-vector space V of dimension $n+1$, identify the K-vector space $V_K \otimes_K g_*V_K$ and the natural descent data on it (see Exercise 2.1.5(iii)) with the space of $(n+1) \times (n+1)$ matrices over K and the K-semilinear involution on it given by the formula

$$M \mapsto g(M)^T\,.$$

Thus, the k-vector space $(V_K \otimes_K g_* V_K)^G$ is identified with the space of Hermitian matrices over K. In turn, the latter space can be identified with the space of all $(n+1) \times (n+1)$ matrices over k as follows: a matrix A over k corresponds to the Hermitian matrix $(A + A^T) + \sqrt{a}(A - A^T)$ over K. Now use Exercise 2.1.5(i) and (ii).

(iii) In the notation of part (ii), show that the surface $R_{K/k}(\mathbb{P}^1)$ is isomorphic to the quadric in $\mathbb{P}^3$ given by the equation

$$x^2 - y^2 + z^2 - au^2 = 0\,.$$

In particular, when $k = \mathbb{R}$ and $a = -1$, we have $K = \mathbb{C}$ and $\mathbb{R}$-points of the real algebraic variety $R_{\mathbb{C}/\mathbb{R}}(\mathbb{P}^1)$ form the Riemann sphere.

(iv) Let $Q \subset \mathbb{P}^3$ be a smooth two-dimensional quadric over k such that $\mathrm{d}(Q) \neq 1$ (see Definition 6.1.1). Let $K = k\big(\sqrt{\mathrm{d}(Q)}\big)$ and let C be a smooth conic over K that corresponds to $\mathrm{cl}(Q_K)$ (see Definition 6.1.3). Prove that $R_{K/k}(C) \cong Q$.

Hint. By Exercise 6.1.2(vi), there is an isomorphism $Q_K \cong C \times C$ of varieties over K. By Exercise 6.2.1(i), the non-trivial involution of K over k permutes the factors in this decomposition.

(v) Use part (iv) to solve Exercise 6.1.4(ii).

8.2. Algebraic tori

In what follows, $\mathbb{G}_m$ denotes the affine group scheme $\mathrm{Spec}(\mathbb{Z}[t, t^{-1}])$ over $\mathbb{Z}$. Given a field F, we denote also by $\mathbb{G}_m$ the corresponding one-dimensional torus over F, that is, the scalar extension of the scheme $\mathbb{G}_m$ from $\mathbb{Z}$ to F.

DEFINITION 8.2.1. An *algebraic torus* over a field k is an algebraic group T over k such that there is a field extension $k \subset F$ and an isomorphism of F-varieties $T_F \cong \mathbb{G}_m^{\times n}$ for a non-negative integer n. In this case, we say that T *splits* over F.

Note that any algebraic torus splits over a finite separable extension of the field of definition (see [**Vos98**, § 3.4]). Thus algebraic tori are the same as forms of $\mathbb{G}_m^{\times n}$.

As above, let $k \subset K$ be a finite separable extension of fields of degree d. The Weil restriction $R_{K/k}(\mathbb{G}_m)$ is an algebraic torus of dimension d over k (this follows from Exercise 8.1.2(ii) and (v)). By Exercise 8.1.2(iv), there is a bijection

$$R_{K/k}(\mathbb{G}_m)(k) \cong K^*\,.$$

For short, we denote this algebraic torus over k by K^*. This notation should not lead to confusion as the field k is fixed from now on. In particular, k^* denotes the torus $\mathbb{G}_m$ over k.

In what follows, K' denotes a finite separable extension of the field K.

EXERCISE 8.2.2. **General properties of algebraic tori**

(i) Prove that the Weil restriction of an algebraic torus is an algebraic torus as well.

(ii) Show that the torus K^* is the twist of the variety $\mathbb{G}_m^{\times d}$ by the homomorphism

$$\varphi\colon G_k \to \mathrm{S}_d \subset \mathrm{Aut}\left(\mathbb{G}_m^{\times d}\right).$$

Hint. Use Exercise 8.1.2(v).

(iii) Prove that the embedding $k^* \hookrightarrow K^*$ is a morphism of algebraic tori.

Hint. Use Exercise 8.1.2(iii).

(iv) Prove that the norm

$$\mathrm{Nm}_{K/k}\colon K^* \to k^*$$

is a morphism of algebraic tori.

Hint. Show that $\mathrm{Nm}_{K/k}$ is the twist of the morphism

$$\mathbb{G}_m^{\times d} \to \mathbb{G}_m\,, \qquad (z_1, \ldots, z_d) \mapsto z_1 \cdot \ldots \cdot z_d$$

by the homomorphism φ from part (ii).

(v) Find equations that define the algebraic torus $R_{K/k}(\mathbb{G}_m)$.

Hint. Consider the norm

$$\mathrm{Nm}_{K/k}\colon K^* \to k^*$$

as a form of degree d over k. It follows from parts (ii) and (iv) that the algebraic torus $R_{K/k}(\mathbb{G}_m)$ is the complement of the corresponding hypersurface in the affine space $\mathbb{A}^d$.

(vi) Prove that the embedding $K^* \hookrightarrow K'^*$ and the norm $\mathrm{Nm}_{K'/K}\colon K'^* \to K^*$ are morphisms of algebraic tori.

Hint. First use parts (iii) and (iv), replacing k and K by K and K', respectively. Then apply Weil restriction from K to k.

EXERCISE 8.2.3. **Rationality of some algebraic tori**

(o) Show that the algebraic torus K^* is rational over k.

Hint. Use Exercise 8.1.5.

(i) Prove that the algebraic torus K^*/k^* is rational over k.

Hint. This algebraic torus is an open subset in the projective space $\mathbb{P}(K) \cong \mathbb{P}^{d-1}$ over k.

(ii) Prove that the algebraic torus K'^*/K^* is rational over k.

Hint. First use part (i), replacing k and K by K and K', respectively. Then apply Weil restriction from K to k and use Exercise 8.1.5.

EXERCISE 8.2.4. **Kernel of the norm**

(i) Prove that the kernel of the norm $\mathrm{Nm}_{K/k}\colon K^* \to k^*$ is a $(d-1)$-dimensional algebraic torus over k.

Hint. Using the hint for Exercise 8.2.2(iv), check that this kernel is isomorphic over k^{sep} to the torus $\mathbb{G}_m^{\times(d-1)}$.

(ii) Prove that the kernel of the norm $\mathrm{Nm}_{K'/K}\colon K'^* \to K^*$ is an algebraic torus over k.

Hint. First use part (i), replacing k and K by K and K', respectively. Then apply Weil restriction from K to k and use Exercise 8.2.2(i).

EXERCISE 8.2.5. **One-dimensional tori**

Let $k \subset K$ be a quadratic separable extension of fields. According to Exercise 8.2.4(i), the kernel of the norm $\mathrm{Nm}_{K/k}\colon K^* \to k^*$ is a one-dimensional algebraic torus over k. Prove that this defines a bijection between the set of quadratic separable extensions of k and the set of one-dimensional tori over k.

Hint. Both sets are bijective with the set of quadratic characters of the Galois group G_k. For the set of one-dimensional algebraic tori this follows from the isomorphism $\mathrm{Aut}(\mathbb{G}_m) \cong \mathbb{Z}/2\mathbb{Z}$.

8.3. Algebraic tori and Galois modules

As above, G_k denotes the Galois group $\mathrm{Gal}\left(k^{sep}/k\right)$.

DEFINITION 8.3.1. Given an algebraic torus T over a field k, the *dual Galois module* over the group G_k is defined by the formula

$$T^\vee = \mathrm{Hom}(T_{k^{sep}}, \mathbb{G}_m)\,.$$

Here, the Galois group G_k acts on $T^\vee$ by the formula

$$({}^g\chi)(t) = g\big(\chi(g^{-1}t)\big),\; g \in G_k, \quad \chi \in T^\vee,\; t \in T(k^{sep})\,.$$

EXERCISE 8.3.2. **Descent data for split tori**

Let M be a free finitely generated abelian group. Consider the split algebraic torus $T = \mathrm{Spec}\left(k[M]\right)$ over the field k, where $k[M]$ is the group algebra of a group M and the group structure on T corresponds to the homomorphism of algebras

$$k[M] \to k[M] \otimes_k k[M], \quad m \mapsto m \otimes m, \quad m \in M\,.$$

(i) Show that there is a canonical isomorphism of abelian groups $M \cong T^\vee$.

(ii) Show that there is a natural isomorphism of rings

$$\mathrm{End}(T) \cong \mathrm{End}(T^\vee)\,,$$

where End on the left-hand side denotes the ring of endomorphisms of the algebraic torus T, and End on the right-hand side denotes the ring of endomorphisms of the abelian group $T^\vee$.

Hint. Using part (i), construct mutually inverse homomorphisms between these rings.

(iii) Let $k \subset K$ be a finite Galois extension. Construct a canonical bijection between the set of descent data (see Example 2.1.2(iv)) on the algebraic torus $\mathrm{Spec}\left(K[M]\right)$ and the set of actions of the Galois group $\mathrm{Gal}(K/k)$ on M.

Hint. Use part (ii) and Exercise 2.2.1(i).

DEFINITION 8.3.3 (cf. Definition 8.3.1). Let M be a G_k-module which is free and finitely generated as an abelian group. The *dual torus* $M^\vee$ of M is the descent from K to k of the algebraic torus $\mathrm{Spec}\left(K[M]\right)$ over K, where $k \subset K$ is a finite Galois extension such that the action of G_k on M factors through its quotient $\mathrm{Gal}(K/k)$; see Exercises 8.3.2(iii) and 2.1.4(vi).

It is easy to check that Definition 8.3.3 makes sense, that is, it does not depend on the choice of the field K.

EXERCISE 8.3.4. **Duality between algebraic tori and Galois modules**

Check that the functors $\Phi\colon T \mapsto T^\vee$ and $\Psi\colon M \mapsto M^\vee$ are quasi-inverse contravariant functors between the category of algebraic tori over k and the category of G_k-modules that are free and finitely generated as abelian groups (that is, check that the functors $\Phi \circ \Psi$ and $\Psi \circ \Phi$ are isomorphic to the identity functors).

It is natural to expect that algebraic tori that are dual to permutation or stably permutation G_k-modules possess some special properties.

DEFINITION 8.3.5. Let T be an algebraic torus over k. One says that T is a *permutation* (respectively, *stably permutation*) torus if $T^\vee$ is a permutation (respectively, stably permutation) G_k-module; see Definitions 1.4.1 and 1.4.3.

EXERCISE 8.3.6. **Structure of permutation tori**

(i) Given a finite separable extension $k \subset K$, construct an isomorphism of G_k-modules
$$(K^*)^\vee \cong \mathbb{Z}[G_k/G_K]\,,$$
where G_k/G_K is the set of cosets.

(ii) Prove that any permutation torus T over k is of the form
$$T \cong \prod_i K_i^*\,,$$
where K_i are some finite separable extensions of k.

Hint. Use part (i), Exercise 1.4.2(i), and Exercise 8.3.4.

EXERCISE 8.3.7. **Triviality of torsors under a stably permutation torus**

(i) Let T be a stably permutation torus over k. Prove that
$$H^1\big(G_k, T(k^{sep})\big) = 0\,.$$

Hint. For a permutation torus, use Exercise 8.3.6(ii), Exercise 8.1.3(vi), Shapiro's lemma from Exercise 1.2.5(iii), and Hilbert's Theorem 90. The case of a stably permutation torus can be reduced easily to the case of a permutation torus.

(ii) Prove that any torsor under a stably permutation torus is trivial.

Hint. Use part (i) and Exercise 2.6.2(i).

EXERCISE 8.3.8. **Stable rationality of stably permutation tori**

(i) Prove that any permutation torus is rational.

Hint. Use Exercises 8.3.6(ii) and 8.2.3(o).

(ii) Using part (i), show that any stably permutation torus is stably rational (see Definition 4.3.3).

EXERCISE 8.3.9. **Torsors under stably permutation tori**

Let T be a stably permutation torus over k, and let X be an irreducible variety over k. Let $\phi\colon V \to X$ be a torsor under T (see Definition 2.6.3). Prove that the varieties V and $X \times T$ are stably birational in the sense of Definition 4.3.3.

Hint. Apply Exercise 1.4.4 and then Exercises 8.3.7(ii), 2.6.4(i), and 8.3.8(ii).

8.4. Universal torsor

In this section, X is a smooth variety over a field k such that the Galois module $\mathrm{Pic}(X_{k^{sep}})$ is free and finitely generated as an abelian group.

DEFINITION 8.4.1. The *Néron–Severi torus* of the variety X is the algebraic torus

$$T_{\mathrm{NS}}(X) = \mathrm{Pic}(X_{k^{sep}})^{\vee}$$

over the field k, where $\mathrm{Pic}(X_{k^{sep}})^{\vee}$ denotes the dual torus of the G_k-module $\mathrm{Pic}(X_{k^{sep}})$ in the sense of Definition 8.3.3.

EXERCISE 8.4.2. **Construction of the universal torsor over $X_{k^{sep}}$**

(i) Let $L_1, \ldots, L_n$ be a collection of line bundles such that their classes $l_1, \ldots, l_n$ form a basis in $\mathrm{Pic}(X_{k^{sep}})$. Let L_i° denote the complement of the zero section in the total space of the bundle L_i. Consider the fibered product

$$W = L_1^{\circ} \times_X L_2^{\circ} \times_X \ldots \times_X L_n^{\circ}\,.$$

Consider the isomorphism

$$T_{\mathrm{NS}}(X)_{k^{sep}} \cong \mathbb{G}_m^{\times n}$$

that is sent by the equivalence of categories from Exercise 8.3.4 to the isomorphism

$$\mathrm{Pic}(X_{k^{sep}}) \cong \bigoplus_{i=1}^{n} \mathbb{Z} \cdot l_i\,.$$

Show that this defines on W a structure of a torsor under $T_{\mathrm{NS}}(X)_{k^{sep}}$ over $X_{k^{sep}}$.

(ii) Prove that the isomorphism class of the $T_{\mathrm{NS}}(X)_{k^{sep}}$-torsor W does not depend on the choice of the line bundles L_i.

DEFINITION 8.4.3. The torsor W constructed in Exercise 8.4.2 is called the *universal torsor* over the variety $X_{k^{sep}}$.

EXERCISE 8.4.4. **Descent for the universal torsor**

Suppose that the variety X has a k-point p and that all invertible functions on $X_{k^{sep}}$ are constant. As in Definition 8.4.3, denote the universal torsor over the variety $X_{k^{sep}}$ by W.

(i) Show that for any element $g \in G_k$, there is a canonical isomorphism

$$g_*W \cong g_*L_1^{\circ} \times_X \ldots \times_X g_*L_n^{\circ}\,,$$

where g_*W and $g_*L_i^{\circ}$ are defined as in Example 2.1.2(v).

(ii) Prove that there is a natural isomorphism between the group of automorphisms of the $T_{\mathrm{NS}}(X)_{k^{sep}}$-torsor W over the variety $X_{k^{sep}}$ and the group of automorphisms of the fiber $W|_p$ of the torsor W over the point p considered as a $T_{\mathrm{NS}}(X)_{k^{sep}}$-torsor over the field k^{sep}.

Hint. Recall that all invertible functions on $X_{k^{sep}}$ are constant.

(iii) Using part (ii), prove that there is a natural bijection between the set of isomorphism classes of descent data on the torsor W over $X_{k^{sep}}$ and the set of isomorphism classes of descent data on the torsor $W|_p$ over k^{sep}.
Hint. Use part (ii) and Exercise 2.2.1(i).
In particular, one has descent data on the torsor W.

(iv) Suppose that
$$H^1\big(G_k, T_{\mathrm{NS}}(X)_{k^{sep}}\big) = \{1\}\,.$$
Prove that in this case there exists a unique $T_{\mathrm{NS}}(X)$-torsor V_X over X such that there is an isomorphism of $T_{\mathrm{NS}}(X)_{k^{sep}}$-torsors
$$\big(V_X\big)_{k^{sep}} \cong W\,.$$

DEFINITION 8.4.5. The torsor V_X constructed in Exercise 8.4.4(iv) is called the *universal torsor* over the variety X.

EXERCISE 8.4.6. **Stable birational equivalence of X and V_X**
Suppose that the variety X has a k-point and that all invertible functions on $X_{k^{sep}}$ are constant. Also, suppose that the G_k-module $\mathrm{Pic}(X_{k^{sep}})$ is stably permutation. Prove that in this case the varieties X and V_X are stably birational.
Hint. Use Exercise 8.3.9.

8.5. Châtelet surfaces and stably permutation modules

The aim of this section is to construct a (discrete) stably permutation Galois module of geometric origin that cannot be obtained as in Exercise 2.5.2. Specifically, we construct a surface S such that its Picard group is a stably permutation Galois module, but it is at least not obvious that the surface is rational, and in general this is even not the case.

Moreover, we already start to approach our next big goal, namely construction of a stably rational but not rational variety (see Chapter 9). Thus we are interested in finding a surface S that has a chance of being stably rational. In particular, S must be unirational at least. Under some mild additional assumptions, this implies that S is rational over the separable closure of the ground field (actually, one does not need any additional assumptions when the ground field is of characteristic zero). Hence we are not satisfied by trivial examples, such as, when S is a surface of general type with Picard group isomorphic to $\mathbb{Z}$.

For the same reasons, we deal with surfaces and are not satisfied, for instance, with higher-dimensional non-rational Fano varieties with Picard group isomorphic to $\mathbb{Z}$. One could take, as such a non-interesting example, a smooth three-dimensional quartic, that is, a smooth hypersurface of degree 4 in $\mathbb{P}^4$. Any such hypersurface is non-rational (see [**IM71**]) and its Picard group is isomorphic to $\mathbb{Z}$; in addition, one has explicit constructions of unirationality for some hypersurfaces of this type. On the other hand, at least a very general three-dimensional quartic is not stably rational (see [**CTP16b**]), while we will be interested in stable rationality in Chapter 9.

We follow the construction of A. Beauville, J.-L. Coilliot-Thélène, J.-J. Sansuc, and P. Swinnerton-Dyer from [**BCTSSD85**, §2].

Let k be a field of characteristic different from 2. As usual, G_k denotes the absolute Galois group $\mathrm{Gal}(k^{sep}/k)$. Choose a separable polynomial P over k. By definition, this means that P does not have multiple roots in the algebraic closure of k. In particular, all roots of P are in k^{sep}. Let $R \subset \mathbb{A}^1$ denote the set of roots of the polynomial P in the field k^{sep}. Until the end of this section, we assume that the degree $d = 2r - 1 \geqslant 1$ of P is odd.

Consider the affine space $\mathbb{A}^3$ over k with coordinates v_1, v_2, and u. Fix an invertible element $a \in k^*$ and define a surface S^0 in $\mathbb{A}^3$ by the equation

$$v_1^2 - av_2^2 = P(u)\,. \tag{8.1}$$

The affine surface S^0 is an open subset of the surface given in $\mathbb{P}^2 \times \mathbb{A}^1$ by the equation

$$v_1^2 - av_2^2 - P(u)v_3^2 = 0\,,$$

where v_i are homogeneous coordinates on $\mathbb{P}^2$ and u is a coordinate on $\mathbb{A}^1$.

Let $\mathbb{P}$ be the projectivization of the vector bundle $\mathcal{O}(r) \oplus \mathcal{O}(r) \oplus \mathcal{O}$ on $\mathbb{P}^1$. Consider an embedding $\mathbb{P}^2 \times \mathbb{A}^1 \hookrightarrow \mathbb{P}$ such that the coordinates v_1 and v_2 on $\mathbb{P}^2$ correspond to the line bundles $\mathcal{O}(r)$ while the coordinate v_3 corresponds to the line bundle $\mathcal{O}$. Let S be the closure of S^0 in $\mathbb{P}$. The surface S is called a *Châtelet surface.* (Note that one usually assumes $d = 3$ (or $d = 4$) when defining a Châtelet surface.) The natural projection $\mathbb{P} \to \mathbb{P}^1$ defines a structure of a conic bundle $f\colon S \to \mathbb{P}^1$ on S. The morphism f is given on S^0 by the formula

$$(v_1, v_2, u) \mapsto u\,.$$

By $F_p \subset S$ we denote the fiber of f at the point $p \in \mathbb{P}^1$; in particular, F_∞ denotes the fiber of f at the point $\infty \in \mathbb{P}^1$.

EXERCISE 8.5.1. **Divisors on the Châtelet surface**

(o) Show that, in suitable coordinates, the surface $S \subset \mathbb{P}$ can be given "at infinity" by the equation

$$V_1^2 - aV_2^2 - Q(U)V_3^2 = 0\,,$$

where

$$Q(U) = U^{2r}P(U^{-1})\,.$$

Show that the surface S is smooth.

Hint. Recall the transition functions for the line bundle $\mathcal{O}(r)$, namely, $V_1 = v_1u^{-r}$. Keep in mind that, according to our conventions, the projectivization of a vector bundle is the variety of lines in fibers.

(i) Check that for any $p \in R \cup \{\infty\}$, the fiber F_p is geometrically reducible and $F_p = D_p \cup D'_p$ for some (different) irreducible divisors D_p and D'_p over the field k^{sep}. Besides them, all other fibers of f are smooth.

(ii) Show that over the field k^{sep} there is a decomposition

$$S \setminus S^0 = F_\infty \cup E \cup E'\,,$$

where the divisors E and E' do not meet each other and are sections of f (in general, these sections are defined over k^{sep}). In what follows, we assume that the divisors E, E', and also D_p, D'_p for all $p \in R \cup \{\infty\}$ are chosen so that the curve E intersects all the curves D_p and does not intersect any of the curves D'_p and, conversely, the curve E' intersects all the curves D'_p and does not intersect any of the curves D_p. Check that one can always make such a choice.

(iii) Check that over the field k^{sep}, there is a linear equivalence of divisors

$$\sum_{p\in R} D_p + D_\infty + E - E' - rF_\infty \sim 0\,.$$

Hint. Consider the function $(v_1 - \sqrt{a}v_2)/v_3$.

(iv) Show that the Picard group $\mathrm{Pic}\left(S_{k^{sep}}\right)$ is freely generated by the divisors E, E', F_∞, and D_p, where $p \in R$.

Hint. Contract over k^{sep} all components of reducible fibers of f and get a $\mathbb{P}^1$-bundle over $\mathbb{P}^1$. Deduce that the Picard group $\mathrm{Pic}\left(S_{k^{sep}}\right)$ is freely generated by the divisors D_∞, E, F_∞, and D_p, where $p \in R$. Then use part (iii).

(v) Show that there is a natural exact sequence of G_k-modules

$$0 \to \mathbb{Z}[R] \to \mathbb{Z}\big[\{D_p, D'_p\}_{p\in R}\big] \oplus \mathbb{Z}\big[\{E, E'\}\big] \oplus \mathbb{Z}\big[\{F_\infty\}\big] \to \mathrm{Pic}\left(S_{k^{sep}}\right) \to 0\,.$$

Hint. Define the first map so that it sends $p \in R$ to the divisor $D_p + D'_p - F_\infty$, and use part (iv).

Recall that for $n > 1$, a *dihedral group* D_n of order $2n$ is the group of symmetries of the plane that preserve a regular n-gon. In particular, there are isomorphisms $\mathrm{D}_2 \cong \mathbb{Z}/2\mathbb{Z} \times \mathbb{Z}/2\mathbb{Z}$ and $\mathrm{D}_3 \cong \mathrm{S}_3$. For any n, there is a (non-canonically split) exact sequence of groups

$$1 \to \mathbb{Z}/n\mathbb{Z} \to \mathrm{D}_n \to \mathbb{Z}/2\mathbb{Z} \to 1\,.$$

Let us say that a polynomial of degree $n > 1$ over k is *dihedral* if it is separable, irreducible over k, and such that the Galois group of its splitting field is isomorphic to the dihedral group D_n. We say that a pair (P, a) is *dihedral* if P is a dihedral polynomial and the extension $k \subset k(\sqrt{a})$ coincides with the quadratic subextension of the splitting field of P that corresponds to the canonical surjection from the dihedral group to $\mathbb{Z}/2\mathbb{Z}$. In particular, if a pair (P, a) is dihedral, then a is not a square in the field k. An example of a dihedral pair is given by a pair (P, a) where P is an irreducible polynomial of degree 3 and a is its discriminant, provided that $a \notin (k^*)^2$.

EXERCISE 8.5.2. **Dihedral pairs**

Suppose that the pair (P, a) is dihedral. Recall that according to our assumptions, the degree d of P is odd.

(i) Let us identify the Galois group G of the splitting field of P with the group D_d, which acts by symmetries of a regular d-gon. Prove that there is a G-equivariant bijection between the set R of roots of P and the set of vertices of the d-gon.

Hint. Since G acts on R transitively, the stabilizer $\mathrm{Stab}_G(\theta)$ of any root $\theta \in R$ has order 2. Since d is odd, the group $\mathrm{Stab}_G(\theta)$ preserves only one vertex v of the d-gon. Now use the fact that the action of D_d on the set of vertices is transitive, and check that the map $\theta \mapsto v$ gives the required bijection.

(ii) Let $\theta \in k^{sep}$ be a root of P. Show that the field $k(\theta, \sqrt{a})$ is the splitting field of P over k.

Hint. Let K be the splitting field of P over k. By the definition of a dihedral pair, the group $\mathrm{Gal}\left(K/k(\sqrt{a})\right)$ is the cyclic subgroup $\mathbb{Z}/d\mathbb{Z}$ of the group $G \cong \mathrm{D}_d$. By Galois theory, the group $\mathrm{Gal}\left(K/k(\theta, \sqrt{a})\right)$ is the stabilizer in the group $\mathbb{Z}/d\mathbb{Z}$ of the root θ. It follows from part (i) that this stabilizer is trivial.

(iii) Let $\theta \in k^{sep}$ be a root of P. Show that over the field $k(\theta)$, there is a decomposition

$$P(u) = \alpha(u - \theta)\left(P_1(u)^2 - aP_2(u)^2\right),$$

where P_1 and P_2 are polynomials of degree $(d-1)/2$ and α is an element of k.

Hint. By part (ii), we know that

$$G \cong \mathrm{Gal}\left(k(\theta, \sqrt{a})/k\right).$$

By part (i), the subgroup

$$\mathrm{Gal}\left(k(\theta, \sqrt{a})/k(\theta)\right) \cong \mathbb{Z}/2\mathbb{Z}$$

of the group G acts freely on the set $R \setminus \{\theta\}$. Consider two different roots

$$\eta = x + \sqrt{a}y, \quad \bar{\eta} = x - \sqrt{a}y, \quad x, y \in k(\theta),$$

of the polynomial P that are interchanged by this action. Then we have

$$(u - \eta)(u - \bar{\eta}) = (u - x)^2 - ay^2.$$

Now by multiplicativity of the norm map for the field extension

$$k(u, \theta) \subset k(u, \theta, \sqrt{a}),$$

we see that over the field $k(\theta)$ the polynomial P has the required form. For α one can take, for instance, the leading coefficient of P.

In what follows, P is an arbitrary separable polynomial over k of odd degree d and $a \in k^*$, as in the beginning of this section. Also, S denotes the corresponding Châtelet surface.

EXERCISE 8.5.3. **Rationality of the Châtelet surface over extensions of k**

(i) Show that over the field $k(\sqrt{a})$ the surface S is rational.

Hint. Check that the sections E and E' are defined over the field $k(\sqrt{a})$.

(ii) Suppose that the pair (P, a) is dihedral. Let $\theta \in k^{sep}$ be a root of P. Prove that over the field $k(\theta)$ the surface S is rational.

Hint. By Exercise 8.5.2(iii), over the field $k(\theta)$ there is a decomposition

$$P(u) = \alpha(u - \theta)\left(P_1(u)^2 - aP_2(u)^2\right).$$

Using multiplicativity of the norm map and making a suitable change of coordinates, rewrite equation (8.1) of the affine open subset $S^0 \subset S$ in the form

$$\hat{v}_1^2 - a\hat{v}_2^2 = \hat{u}.$$

Now the projection of S^0 to the plane $\mathbb{A}^2$ with coordinates $\hat{v}_1$ and $\hat{v}_2$ is an isomorphism.

Exercise 8.5.3 shows that the Châtelet surface is rational over many simply constructed finite extensions of the field k. However, it turns out that over the field k itself such surfaces are often non-rational. In order to show this, we use a theorem of V. A. Iskovskikh on non-rationality of conic bundles (see [**Isk67**, Theorem 1.6], [**Isk70**, Theorem 1], and [**Isk71a**, Theorem 2]; a slightly more modern exposition can be found in [**Isk96**, §4]).

Suppose that the field k is perfect. Let Σ be a smooth projective surface and $f\colon \Sigma \to \mathbb{P}^1$ a conic bundle over k. Suppose that the conic bundle is *relatively minimal* over k, that is, there is no G_k-invariant collection of pairwise non-intersecting (-1)-curves on Σ over the field k^{sep} that are contracted by the map f. *Iskovskikh's theorem* states that in this case, the surface Σ is rational over k if and only if there exists a k-point on Σ and the number of geometrically reducible fibers of f over k^{sep} is at most three. (Note that smoothness of the surface Σ implies that all fibers of the morphism f are reduced!)

In particular, Iskovskikh's theorem gives another (more geometric) solution to Exercise 8.5.3(ii). Indeed, over the field $k(\theta)$ one can contract components of geometrically reducible fibers of f so that the conic bundle obtained (defined over $k(\theta)$) has at most two geometrically reducible fibers, and each of these fibers is defined over $k(\theta)$. Also, the corresponding surface has a point over the field $k(\theta)$, which implies its rationality.

EXERCISE 8.5.4. **Non-rationality of the Châtelet surface over the field k**

(i) Prove that the conic bundle $f\colon S \to \mathbb{P}^1$ is relatively minimal over k if and only if for any root θ of P, the field $k(\theta)$ does not contain $\sqrt{a}$.

Hint. First suppose that for any θ, the field $k(\theta)$ does not contain $\sqrt{a}$. Then for any geometrically reducible fiber F_p of f, there is an element in the group G_k that preserves the point $p \in \mathbb{P}^1$ and permutes the components of this fiber. This means that the surface S is relatively minimal over k. Now suppose that for some root θ, the field $k(\theta)$ does not contain $\sqrt{a}$. Check that the G_k-orbit of any of the components of the fiber F_θ consists of two disjoint (-1)-curves.

(ii) Suppose that the field k is perfect. Suppose that the degree of the polynomial P is at least 3 and that for any root θ of P the field $k(\theta)$ does not contain $\sqrt{a}$. Show that in this case, the Châtelet surface S is non-rational over k.

Hint. Use part (i) and Iskovskikh's theorem.

In particular, this holds when the polynomial P is irreducible, because, by our assumption, its degree is odd.

The goal of the following exercise is to show that for dihedral pairs, the Picard group $\operatorname{Pic}\left(S_{k^{sep}}\right)$ is a stably permutation G_k-torus. We will use a result of Endo and Miyata (see [**EM75**, Theorem 1.5] and [**EM75**, Lemma 1.2]). Let G be a finite group such that all its Sylow subgroups are cyclic. Let

$$0 \to M' \to M \to M'' \to 0$$

be an exact sequence of G-modules, where M' and M'' satisfy condition (1.3) from Exercise 1.4.5(ii) and are free and finitely generated as abelian groups. The *Endo–Miyata theorem* states that any such exact sequence splits.

EXERCISE 8.5.5. **Picard group and universal torsor for the Châtelet surface**

Suppose that the pair (P, a) is dihedral. Let $G \cong \mathrm{D}_d$ be the Galois group of the splitting field of the polynomial P.

(o) Check that the action of the group G_k on all modules in the exact sequence from Exercise 8.5.1(v) factors through its quotient G.

(i) Let $H \subset G$ be a cyclic subgroup of order d. Show that the group $\mathrm{Pic}\left(S_{k^{sep}}\right)$ is a stably permutation H-module.

Hint. The surface S is rational over the field $k(\sqrt{a})$ by Exercise 8.5.3(i). Now one can use Exercise 2.5.2(iv). Note that in the case of surfaces, the assumption from Exercise 2.5.2 that the characteristic of k is zero is not necessary, because in dimension 2 one has resolution of singularities over an arbitrary field.

(ii) Let $\theta \in k^{sep}$ be a root of the polynomial P. Consider its stabilizer

$$\mathbb{Z}/2\mathbb{Z} \cong H_\theta \subset G\,.$$

Show that $\mathrm{Pic}\left(S_{k^{sep}}\right)$ is a stably permutation H_θ-module.

Hint. The surface S is rational over the field $k(\theta)$ by Exercise 8.5.3(ii), so that one can use Exercise 2.5.2(iv).

(iii) Let l be a prime number, and let $G_l \subset G$ be a Sylow l-subgroup. Check that $\mathrm{Pic}\left(S_{k^{sep}}\right)$ is a stably permutation G_l-module.

Hint. Since the degree of the polynomial P is odd, every non-trivial 2-subgroup of G is one of the groups H_θ. Hence for $l = 2$ the assertion follows from part (ii). Thus, one can assume that l is odd. Then G_l is contained in the subgroup H. In this case, the assertion follows from part (i) and Exercise 1.4.4.

(iv) Show that $\mathrm{Pic}\left(S_{k^{sep}}\right)$ is a stably permutation G-module, hence a stably permutation G_k-module.

Hint. Apply the Endo–Miyata theorem to the exact sequence from Exercise 8.5.1(v), using part (iii) and Exercise 1.4.5(iii) to check all assumptions of the theorem.

(v) Prove that there exists a universal torsor V_S of S and that S and V_S are stably birational.

Hint. The surface S is projective, so all invertible functions on it are constant. Besides that, the surface S has a k-point, which can be found easily on the geometrically reducible fiber D_∞ of the conic bundle f. Thus a universal torsor V_S exists by Exercise 8.4.4(iv). The fact that S and V_S are stably birational follows from Exercise 8.4.6.

8.6. Further reading

One finds, for instance, in [**BLR90**, §7.6] the definition of Weil restriction for schemes and its general properties. In [**Tab15**] Weil restriction is investigated in the non-commutative context.

One of the best references for algebraic tori is the excellent book [**Vos98**], which also contains a lot of material on birational properties of more general algebraic groups. A detailed survey of results on rationality questions for algebraic tori, as well as on many related arithmetic questions, is given in [**Kun07**]; see also references therein.

Universal torsors were studied systematically in the foundational paper [**CTS87**]. Exercise 8.4.6 shows that it is useful to have a detailed description of universal torsors. This problem has been addressed in many papers; in particular, universal torsors were described explicitly for Châtelet surfaces in [**CTSSD87a**] and [**CTSSD87b**], and for many other rational surfaces by using Cox rings in [**HT04**].

Let us also mention the paper [**dlBBP12**] about rational points on Châtelet surfaces.

CHAPTER 9

Example of a Non-rational Stably Rational Variety

9.1. Plan of the construction

In this chapter, following a paper of A. Beauville, J.-L. Colliot-Thélène, J.-J. Sansuc, and P. Swinnerton-Dyer, [**BCTSSD85**], we construct a variety X over a perfect field k of characteristic different from 2 such that X is non-rational but is stably rational (and, in particular, is unirational). If k is algebraically closed, then, clearly, X cannot be a curve. Actually, this is also true when k is an arbitrary field (this follows from the fact that a conic without points is not unirational). Castelnuovo's theorem implies that over an algebraically closed field of characteristic zero any unirational surface is rational (see [**Bea96**, Corollary V.5]), so that, in this case, X cannot be a surface either. Thus it is natural to look for an example among non-rational surfaces over a non-algebraically closed field k that are rational over the algebraic closure $\bar{k}$ of k. One of the simplest examples of surfaces of this type can be found among conic bundles over $\mathbb{P}^1$ with the help of Iskovskikh's theorem from §8.5. More precisely, we will work with quasi-projective surfaces and X will be an open subset in a certain non-rational (projective) conic bundle $\widetilde{X}$ over $\mathbb{P}^1$.

The results from §8.5 suggest a guideline for where to find the needed example. Specifically, we would like to construct a projective surface $\widetilde{X}$ which is non-rational but stably rational. According to Exercise 2.5.2(v), in this case, the Picard group $\mathrm{Pic}\left(\widetilde{X}_{k^{sep}}\right)$ a stably permutation G_k-module, which restricts our search significantly. Moreover, we already have a series of examples with this property: these are (many of) the Châtelet surfaces, as we know from Exercises 8.5.4(ii) and 8.5.5(iv). Furthermore, since the G_k-module $\mathrm{Pic}\left(\widetilde{X}_{k^{sep}}\right)$ is stably permutation, it follows from Exercise 8.4.6 (see also Exercise 8.5.5(v)) that the surface $\widetilde{X}$ is stably birational to its universal torsor $V_{\widetilde{X}}$. Thus, in order to solve the initial problem, it is enough to establish stable rationality for the universal torsor of a suitable Châtelet surface (in fact, we will check its rationality, not just stable rationality). It turns out that such torsors can be given by explicit equations (see [**CTSSD87b**] and [**CTSSD87c**]). The question of their rationality is reduced to the question of rationality of intersections of two quadrics.

It will be convenient for us to work not with the universal torsor $V_{\widetilde{X}}$ itself, but with its quotient by a certain action of an algebraic torus. The quotient is also rational and, moreover, is also stably birational to X (this quotient will be denoted by V, in order to remind the reader about its relation with the universal torsor). A reason for this replacement is that it decreases the dimension. We will omit intermediate steps that concern the universal torsor and get directly that $X \times \mathbb{P}^3$ is birational to an intersection of two quadrics in $\mathbb{P}^7$. Rationality of this intersection will be obtained by projecting from a line.

Note that when k is algebraically closed, one can construct an example of a three-dimensional non-rational stably rational variety. The construction is based on the two-dimensional example over a non-algebraically closed field; non-rationality is obtained by using the theory of the intermediate Jacobian (for details see [**BCTSSD85**, §3]).

We have explained above where to look for the required example. Now let us give some more details about the plan for the construction of the example (as was already mentioned, the construction will not explicitly involve the universal torsor, but it would be helpful to keep it in mind for a better understanding of the general picture). We will consider the following commutative diagram of varieties:

(9.1)

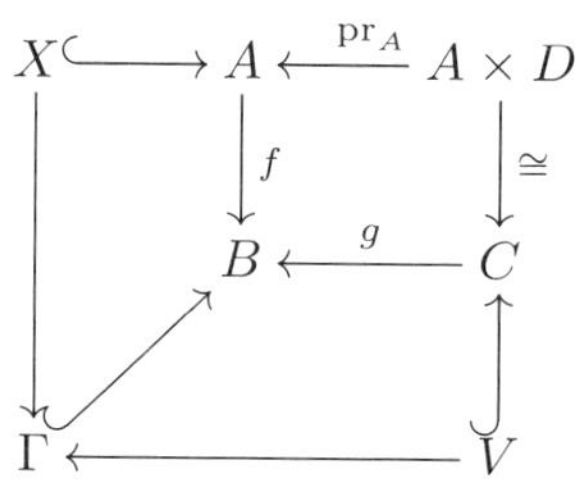

Here Γ is a rational curve on a variety B. We have $X = f^{-1}(\Gamma)$ and $V = g^{-1}(\Gamma)$. The variety X is a non-rational irreducible surface, while V and D are rational varieties. Since $C \cong A \times D$, we have $V \cong X \times D$. Thus X gives the required example. Moreover, in diagram (9.1) the varieties A, B, C, and D are algebraic tori, and all morphisms are morphisms of algebraic tori. After suitable compactifications, f becomes a restriction of a conic bundle, g becomes a restriction of a quadratic rational map from $\mathbb{P}^7$ to $\mathbb{P}^3$, and Γ becomes an open subset of a projective line in $\mathbb{P}^3$. This allows us to consider V as an open subset in an intersection of two quadrics.

9.2. The fields K, k', and K'

In what follows, we will suppose that the ground field k is perfect and its characteristic is different from 2. Note that perfectness of k is needed only to apply Iskovskikh's theorem in the proof of non-rationality of the surface X, and in all other places we can omit this condition, assuming only that all extensions of k that we consider are separable. The assumption on the characteristic of k is needed to prove rationality of the variety V, and more precisely, to work effectively with quadrics that define a compactification of V.

Let $P \in k[x]$ be an irreducible polynomial of degree 3 over k with discriminant a. Suppose that $a \notin (k^*)^2$. Consider the fields $k' = k(\sqrt{a})$ and $K = k(\theta)$, where $\theta \in \bar{k}$ is a root of P. Let K' be the composite of the extensions K and k' of the field k, that is, K' is the splitting field of the polynomial P.

EXERCISE 9.2.1. **Galois groups**
Prove that there are isomorphisms

$$\mathrm{Gal}(K'/k') \cong \mathbb{Z}/3\mathbb{Z}, \quad \mathrm{Gal}(K'/K) \cong \mathrm{Gal}(k'/k) \cong \mathbb{Z}/2\mathbb{Z}, \quad \mathrm{Gal}(K'/k) \cong \mathrm{S}_3 .$$

In particular, the pair (P, a) is dihedral (cf. the example before Exercise 8.5.2). Check that the extension $k \subset K$ is not Galois.

As in Chapter 8, in what follows we denote by K^* both the algebraic torus $R_{K/k}(\mathbb{G}_m)$ and the multiplicative group of the field K (which is canonically isomorphic to the group of k-points of this torus). Analogously, we denote by K both the algebraic group $R_{K/k}(\mathbb{G}_a)$ and the additive group of the field K. The same notation is used for other fields over k (including k itself); in particular, the symbol k denotes also the algebraic group $\mathbb{G}_a = \mathrm{Spec}(k[t])$.

9.3. Non-rational conic bundle

Consider morphisms of algebraic tori $\mathrm{Nm}_{K/k}\colon K^* \to k^*$ and $\mathrm{Nm}_{k'/k}\colon k'^* \to k^*$. Let $A = K^* \times_{k^*} k'^*$ be the fibered product of K^* and k'^* over k^*, and put $B = K^*$. Let $f\colon A \to B$ be the natural projection.

EXERCISE 9.3.1. **The variety A in an explicit form**

Show that the variety A is an open subset in the affine hypersurface in $\mathbb{A}^5$ given by the equation

$$F_3(u_1, u_2, u_3) = v_1^2 - av_2^2\,,$$

where u_1, u_2, u_3, v_1, and v_2 are coordinates in $\mathbb{A}^5$ and F_3 is a homogeneous cubic polynomial.

Hint. The polynomial F_3 is the one that gives the norm $\mathrm{Nm}_{K/k}$.

The element $\theta \in K$ defines a k-point of the algebraic group K over k. The translation by $-\theta$ in the group K applied to the subgroup $k \subset K$ defines a subvariety $k - \theta$ in K (this is an affine line in the affine three-dimensional space over k). The variety K contains K^* as a Zariski-open subset. Consider the algebraic curve

$$\Gamma = (k - \theta) \cap K^* \subset K\,. \tag{9.2}$$

Put $X = f^{-1}(\Gamma) \subset A$.

EXERCISE 9.3.2. **Non-rationality of X**

(i) Show that Γ is isomorphic to the complement in $\mathbb{A}^1$ of the closed subset given by the polynomial P.

Hint. By Exercise 8.2.2(v) and the hint for its solution, the variety K^* is the complement in K of the hypersurface $\{\mathrm{Nm}_{K/k} = 0\}$. Show that the restriction of the norm $\mathrm{Nm}_{K/k}$ to $k - \theta \subset K$ is equal to $P(t)$, where t is a coordinate on $k \cong \mathbb{G}_a$. This can be done, for example, by considering the scalar extension from k to k^{sep} (or to K') and using the hint for Exercise 8.2.2(iv).

(ii) Show that the variety X is an open subset in the affine surface in $\mathbb{A}^3$ given by the equation

$$P(u) = v_1^2 - av_2^2\,,$$

where u, v_1, and v_2 are coordinates on $\mathbb{A}^3$. Thus X is an open subset in the Châtelet surface (see Chapter 8).

(iii) Prove that the surface X is non-rational.

Hint. Use Exercises 8.5.4(ii) and 9.2.1. Note that in our case the degree of the polynomial P is three, that is, the corresponding Châtelet surface has exactly four geometrically reducible fibers over $\mathbb{P}^1$. Therefore, instead of Iskovskikh's theorem in its full generality as used in Exercise 8.5.4(ii), it is enough to use the result from [**Isk71a**].

9.4. Rational intersection of two quadrics

We start this section with a geometric construction that establishes the rationality of a (sufficiently nice) intersection of quadrics containing a line. Recall that a variety $C \subset \mathbb{P}^n$ is a *cone with a vertex at a point* $v \in C$ if for any point $c \in C$ different from v, the line $\langle v, c\rangle$ is contained in C. One can easily see that the set of all vertices of a given cone form a projective subspace in $\mathbb{P}^n$, which is also often called a vertex of the cone (fortunately, it is usually clear from the context what is actually meant by a vertex).

Suppose that a variety $M \subset \mathbb{P}^n$ is a complete intersection of two quadrics Q_1 and Q_2 defined over k. In other words, we have that $\dim(M) = n-2$ and the homogeneous ideal of M is generated by the equations of the quadrics Q_1 and Q_2. In particular, for any point $p \in M$, there is an equality

$$\mathbb{T}_p(M) = \mathbb{T}_p(Q_1) \cap \mathbb{T}_p(Q_2)\,,$$

where $\mathbb{T}_p(-)$ denotes the embedded projective tangent space to a corresponding projective variety at the point p. Suppose that the variety M is irreducible and contains a line l over k such that the following two conditions are satisfied:

(A) the line l is not contained in the set $\mathrm{Sing}(M)$ of singular points of the variety M;

(B) the variety M is not a cone with a vertex being a point on the line l.

The goal of the next exercise is to prove rationality of the variety M.

EXERCISE 9.4.1. **Rationality of an intersection of two quadrics containing a line**

Let $\Pi \subset \mathbb{P}^n$ be a general two-dimensional plane over $\bar{k}$ that passes through the line l. Denote by q_i the restrictions of the quadrics Q_i to the plane Π. Note that generality of Π implies that the q_i do not coincide with Π, that is, they are (possibly, reducible or not reduced) conics on Π containing l.

(o) Show that the variety M is geometrically irreducible and is not contained in any hyperplane. (This information will not be used in the proof below, but it allows one to see the geometric picture better.)

Hint. Since the line l is defined over the field k and the variety M is irreducible, the line l is contained in each irreducible component of the variety $M_{\bar{k}}$. Condition (A) tells us that $l \not\subset \mathrm{Sing}(M)$ and hence such a component is unique, that is, M is geometrically irreducible. Now suppose that M is contained in a hyperplane $H \subset \mathbb{P}^n$. Then M is contained in each of the intersections

$$\hat{Q}_i = Q_i \cap H, \quad i = 1, 2\,.$$

Note that none of these intersections coincides with the hyperplane H, because otherwise either M would be reducible or M would coincide with H (in the latter case M would not be a complete intersection). Since

$$\dim(M) = n-2 = \dim(\hat{Q}_i)\,,$$

the variety M is a common irreducible component of $\hat{Q}_1$ and $\hat{Q}_2$. In particular, one has

$$\deg(M) \leqslant \deg(\hat{Q}_1) \leqslant 2 < 4 = \deg(Q_1)\cdot\deg(Q_2)\,,$$

which is impossible, because M is a complete intersection of Q_1 and Q_2.

(i) Show that $q_i \neq 2l$ for $i = 1, 2$.

Hint. If $q_i = 2l$ for some $i = 1, 2$, then the quadric Q_i is singular at every point of the line l. To check this, consider the tangent space to Q_i at the corresponding point. By assumption, the tangent space contains a general plane passing through the line l. In this case also $l \subset \mathrm{Sing}(M)$, which contradicts condition (A).

Thus one has $q_i = l \cup l_i$, where $l_i \neq l$.

(ii) Show that the lines l_1 and l_2 from part (i) are different.

Hint. Otherwise the quadrics Q_1 and Q_2 would coincide.

(iii) Let $l' \subset l$ be the (non-empty) open subset formed by all points $p \in l$ such that $p \notin \mathrm{Sing}(M)$. Show that the lines l_1 and l_2 cannot intersect at a point $p \in l'$.

Hint. Assume the contrary. Show that the plane Π is contained in the tangent space $\mathbb{T}_p(M) \cong \mathbb{P}^{n-2}$. Furthermore, show that for a general plane Π this inclusion is impossible, providing an upper bound on the dimension of the variety of all planes contained in $\bigcup_{x \in l'} \mathbb{T}_x(M)$.

(iv) Show that the lines l_1 and l_2 intersect at a (unique) point p that does not belong to the line l.

Hint. Suppose that $p \in l$. It follows from part (iii) that p belongs to the finite set $l \setminus l'$. By irreducibility of the variety of planes passing through l, we see that the point p is the same for all planes Π. Considering tangent spaces to the quadrics Q_i at the point p, we see that both quadrics Q_i are singular at p and hence they are cones with vertex at p. This implies that the variety M is also a cone with vertex at $p \in l$, which contradicts condition (B).

(v) Prove that the variety M is rational over k.

Hint. Consider the projection

$$M \dashrightarrow \mathbb{P}^{n-2}$$

of the variety M with center at the line l. Using parts (i), (ii), and (iv), show that this is a birational map.

(vi) Where did we use irreducibility of M in the above argument?

Hint. Actually, it follows from parts (i), (ii), and (iv) only that one of the irreducible components of M is mapped birationally to its image under the projection from l. Without the irreducibility assumption it could be possible that there is another irreducible component whose dimension decreases under the projection from l. For example, if the quadric

$$Q_1 = H \cup H'$$

is reducible, the quadric Q_2 is general, and l is a general line in $Q_2 \cap H$, then the projection from l maps the irreducible component $Q_2 \cap H'$ of the variety M birationally to its image, while the irreducible component $Q_2 \cap H$ is birationally a $\mathbb{P}^1$-bundle over its image.

Now let us come back to the construction of our example. Consider the algebraic torus

$$\hat{C} = K'^* \times k'^*$$

and the morphism of algebraic tori

$$\hat{g}\colon \hat{C} \to B, \quad (x, y) \mapsto \mathrm{Nm}_{K'/K}(x) \cdot \mathrm{Nm}_{k'/k}(y)^{-1}.$$

Put

$$C = \hat{C}/k^*,$$

where k^* is embedded diagonally into $K'^* \times k'^*$. The morphism of algebraic tori $\hat{g}$ factors through a morphism of algebraic tori

$$g\colon C \to B.$$

EXERCISE 9.4.2. **Irreducibility of fibers of g**

(i) Prove that there is an exact sequence of algebraic groups

$$1 \to \mathrm{Ker}(\mathrm{Nm}_{K'/K}) \times \mathrm{Ker}(\mathrm{Nm}_{k'/k}) \to \mathrm{Ker}(\hat{g}) \to k^* \to 1.$$

(ii) Deduce from part (i) that the variety $\mathrm{Ker}(\hat{g})$ is irreducible.

Hint. Use Exercise 8.2.4 and the fact that a variety fibered over an irreducible base with geometrically irreducible fibers is irreducible itself.

(iii) Prove that the kernel of the morphism g is irreducible.

Hint. There is a surjective morphism $\mathrm{Ker}(\hat{g}) \to \mathrm{Ker}(g)$.

(iv) Prove that all fibers of the morphism g are irreducible.

Put $V = g^{-1}(\Gamma) \subset C$, where Γ is the curve given by (9.2). Recall that $\theta \in \bar{k}$ is a root of the polynomial P.

EXERCISE 9.4.3. **Compactification of V**

(i) Consider B as an open subset in the projective space $\mathbb{P}(K \oplus k) \cong \mathbb{P}^3$, and consider C as an open subset in the projective space $\mathbb{P}(K' \oplus k') \cong \mathbb{P}^7$. Show that the morphism g extends to a quadratic rational map

$$\tilde{g}\colon \mathbb{P}(K' \oplus k') \dashrightarrow \mathbb{P}(K \oplus k), \quad (x : y) \mapsto \big(\mathrm{Nm}_{K'/K}(x) : \mathrm{Nm}_{k'/k}(y)\big).$$

(ii) Show that V is an open dense subset in the projective variety

$$\widetilde{V} = \{(x : y) \in \mathbb{P}(K' \oplus k') \mid \pi\big(\mathrm{Nm}_{K'/K}(x) + \theta \cdot \mathrm{Nm}_{k'/k}(y)\big) = 0\},$$

where $\pi\colon K \to K/k$ is the natural projection.

Thus, the variety $\widetilde{V}$ from Exercise 9.4.3(ii) is a compactification of the variety V. It follows from Exercise 9.4.2(iv) that the variety V is irreducible (and so the variety $\widetilde{V}$ is irreducible as well). The goal of the next exercise is to prove that the variety $\widetilde{V}$ does not have singular points defined over k.

EXERCISE 9.4.4. **Singularities of $\widetilde{V}$**

(i) Let $E \subset F$ be a separable quadratic extension of a field E with the non-trivial automorphism $\sigma \in \mathrm{Gal}(F/E)$. Consider E and F as affine spaces over E. Let $N\colon F \to E$ be the morphism given by the formula

$$N(s) = \mathrm{Nm}_{F/E}(s)\,.$$

Let us identify the tangent space to F at a point $s \in F$ with the space F itself, and let us identify the tangent space to E at a point $t \in E$ with the space E. Show that the differential of the morphism N at a point s is

$$dN\colon \alpha \mapsto \mathrm{Tr}_{F/E}\big(\alpha \cdot \sigma(s)\big), \quad \alpha \in F\,,$$

where $\mathrm{Tr}_{F/E}\colon F \to E$ denotes the trace in the separable extension $E \subset F$.

Hint. Use the equality

$$\mathrm{Nm}_{F/E}\big(s\big) = s \cdot \sigma(s)$$

and differentiate it.

(ii) For every point $(x, y) \in K' \oplus k'$, consider the map

$$T_{x,y}\colon (\alpha, \beta) \mapsto \pi\Big(\mathrm{Tr}_{K'/K}\big(\alpha \cdot \sigma(x)\big) + \theta \cdot \mathrm{Tr}_{k'/k}\big(\beta \cdot \sigma(y)\big)\Big), \quad (\alpha, \beta) \in K' \oplus k'\,.$$

Show that a point $(x : y) \in \widetilde{V}(k)$ is singular if and only if the map

$$T_{x,y}\colon K' \oplus k' \to K/k$$

is not surjective.

Hint. Use the equation for $\widetilde{V}$ from Exercise 9.4.3(ii).

(iii) Show that for any point $(x : y) \in \widetilde{V}(k)$, we have $x \neq 0$.

Hint. If $x = 0$, then it follows from Exercise 9.4.3(ii) that

$$\theta \cdot \mathrm{Nm}_{k'/k}(y) \in k\,,$$

which is impossible, because $\theta \notin k$. Note that this is the only place in the whole exercise where one uses any information on the number θ!

(iv) Prove that all k-points of the variety $\widetilde{V}$ are non-singular.

Hint. The map from K' to K given by the formula

(9.3) $$z \mapsto \mathrm{Tr}_{K'/K}\big(z \cdot \sigma(x)\big)$$

is surjective, because the extension $K \subset K'$ is separable and $x \neq 0$ by part (iii).

(v) Is it possible to show similarly that all $\bar{k}$-points on $\widetilde{V}$ are non-singular?

Hint. This is impossible. Indeed, we can extend the definition of the map (9.3) to a map from $\bar{k}^6$ to $\bar{k}^3$, using Weil restriction of scalars. However, the new map is no longer a trace for a field extension in any sense. In particular, it is impossible to use non-degeneracy of the trace, which is essential for part (iv). In fact, as shown in [**BCTSSD85**], the variety $\widetilde{V}$ has eight isolated singular points over the field $\bar{k}$.

One can deduce from Exercise 9.4.4 the needed information on the structure of the variety $\widetilde{V}$ in order to prove its rationality using Exercise 9.4.1. The point $p = \mathbb{P}(k \oplus 0)$ is contained in the closure of the curve $\Gamma \subset K^*$ (see (9.2)) in the projective space $\mathbb{P}(K \oplus k)$. It follows that the variety $\widetilde{V}$ contains the line

$$l = \mathbb{P}(k' \oplus 0) \subset \mathbb{P}(K' \oplus k')\,. \tag{9.4}$$

EXERCISE 9.4.5. **Rationality of V**

(o) Show that the variety $\widetilde{V}$ is an intersection of two quadrics in $\mathbb{P}^7$.

Hint. Use Exercise 9.4.3(ii).

(i) Show that the line l is not contained in the set of singular points of $\widetilde{V}$.

Hint. By Exercise 9.4.4(iv), the variety $\widetilde{V}$ does not have singular k-points, while the line l is defined over k.

(ii) Show that $\widetilde{V}$ is not a cone.

Hint. The vertex of the cone is a projective subspace defined over k. In particular, it has a k-point, which is a singular point on the cone.

(iii) Prove that $\widetilde{V}$ is rational (and hence V is rational as well).

Hint. Use irreducibility of $\widetilde{V}$, parts (i) and (ii), and Exercise 9.4.1.

9.5. Stable birational equivalence between X and V

Our final goal is to extend the morphisms f and g to a commutative diagram (9.1) by choosing a suitable torus D. First we show that g factors through f. For simplicity, we will denote elements of the torus $\hat{C}$ and their images in $C = \hat{C}/k^*$ by the same symbols.

EXERCISE 9.5.1. **The morphism $h\colon C \to A$**

(i) Construct a morphism $h'\colon C \to k'^*$ that satisfies

$$\mathrm{Nm}_{K/k} \circ g = \mathrm{Nm}_{k'/k} \circ h'\,.$$

Hint. Put

$$h'(x, y) = \mathrm{Nm}_{K'/k'}(x) \cdot \mathrm{Nm}_{k'/k}(y)^{-1} \cdot y^{-1}\,,$$

where $x \in K'$ and $y \in k'^*$.

(ii) Show that the morphisms g and h' give a morphism

$$h\colon C \to A = K^* \times_{k^*} k'^*$$

such that $g = f \circ h$.

(iii) Let τ be a generator of the Galois group $\mathrm{Gal}(K'/k') \cong \mathbb{Z}/3\mathbb{Z}$. Define a morphism

$$\lambda\colon A \to C$$

by the formula

$$\lambda(r, s) = \left(\tau(r)^{-1}, s^{-1}\right).$$

Check that λ is a section of the morphism h, that is,

$$h \circ \lambda = \mathrm{id}_A\,.$$

Hint. Let σ denote the non-trivial involution of the field K' over K. Use the relation

$$\sigma\tau = \tau^2\sigma$$

in the group $\mathrm{Gal}(K'/k)$, and also the equality

$$\mathrm{Nm}_{K/k}(z) = \mathrm{Nm}_{K'/k'}(z)$$

for every $z \in K^*$.

The goal of the next exercise is to analyze how far the composition $\lambda \circ h$ is from the identity morphism id_C.

EXERCISE 9.5.2. **The torus D**

(i) Show that for $(x, y) \in C$, the element

$$(x, y) \cdot (\lambda \circ h)(x, y)^{-1} \in C$$

depends only on $x \in K'^*$. This defines a morphism of algebraic tori $K'^* \to C$. Prove that $\tau(K^*) \subset K'^*$ is contained in its kernel.

Hint. Use the explicit formula

$$(x, y) \cdot (\lambda \circ h)(x, y)^{-1} = \Big(x \cdot \tau\big(\mathrm{Nm}_{K'/K}(x)\big), \mathrm{Nm}_{K'/k'}(x)\Big) \in C\,,$$

and also the hint for Exercise 9.5.1(iii).

(ii) Put $D = K'^*/\tau(K^*)$. Show that the torus D is rational.

Hint. Use the isomorphism

$$K'^*/\tau(K^*) \xrightarrow{\tau^{-1}} K'^*/K^*$$

and Exercise 8.2.3(ii).

Exercise 9.5.2(i) provides a morphism $\mu\colon D \to C$.

EXERCISE 9.5.3. **The decomposition $A \times D \cong C$**

(i) Using the morphisms λ and μ from Exercises 9.5.1 and 9.5.2, construct an isomorphism

$$A \times D \xrightarrow{\sim} C\,.$$

Hint. The inverse morphism $C \to A \times D$ is given by the formula

$$(x, y) \mapsto \big(h(x, y), x\big)\,,$$

where $x \in K'^*$ and $y \in k'^*$.

(ii) Check that the diagram (9.1) is commutative.

(iii) Check that the variety V is birational to $X \times D$. In particular, the varieties X and V are stably birational.

Hint. Use part (ii) and Exercise 9.5.2(ii).

Therefore, our initial plan is entirely completed. In particular, we have proved in Exercise 9.3.2(iii) that the surface X is non-rational, and we have checked in Exercise 9.4.5(iii) that the variety V is rational. Finally, we have shown in Exercise 9.5.3(iii) that X and V are stably birational. In other words, the surface X is an example of a non-rational stably rational variety.

9.6. One more construction of stable rationality

N. Shepherd-Barron noticed that the above result can be slightly strengthened: it turns out that already the variety $X \times \mathbb{P}^2$ is rational. Let us explain this using the above notation. We will also need *Voskresenskii's theorem* (see [**Vos67**]), which states that any two-dimensional algebraic torus is rational.

As was shown in Exercise 9.5.3(i), there is an isomorphism of algebraic tori $C \cong A \times D$. At the same time, the torus D contains a one-dimensional subtorus $T = k'^*/k^*$. It follows from Exercise 9.5.2 that the T-orbit of a point $(x, y) \in C$ is the image of the map

$$T \to C\,, \quad t \mapsto (x \cdot t \cdot \mathrm{Nm}_{k'/k}(t), y \cdot t^3)\,, \quad t \in k'^*\,.$$

Since this map has degree 3 with respect to t, the closures of the orbits of T are (rational) cubic curves in the projective space $\mathbb{P}^7 = \mathbb{P}(K' \oplus k')$. The torus T is embedded as an open subset into the projective line $\mathbb{P}^1 = \mathbb{P}(k')$, and the complement is the subvariety $Z \subset \mathbb{P}(k')$ given by the quadratic equation $\mathrm{Nm}_{k'/k}(t) = 0$, where $t \in k'$ (thus, Z is a pair of points over the algebraic closure of the field k). This implies that for all $x \in K'$ and $y \in k'^*$, the boundary of the T-orbit of the point $(x : y) \in \mathbb{P}^7$ is equal to

$$Z \subset \mathbb{P}(k' \oplus 0) = l \subset \mathbb{P}^7\,;$$

cf. (9.4). To prove this, one can check that the conditions $\mathrm{Nm}_{k'/k}(t) = 0$, $t \neq 0$, and $y \neq 0$ imply that $\mathrm{Nm}_{k'/k}(y \cdot t^3) = 0$ and $y \cdot t^3 \neq 0$. Therefore, every hyperplane H that contains the line l intersects such a T-orbit exactly at one point.

Since the diagram (9.1) is commutative, we conclude that the subvariety $V \subset C$ is invariant under the action of the torus T. The previous discussion implies that the section of V by a general hyperplane H that passes through the line l is birational to the quotient V/T (such a section is called a *slice* in geometric invariant theory). The variety $V \cap H$ is still an intersection of two quadrics containing a line that satisfies the conditions of Exercise 9.4.1. Consequently, the variety $V \cap H$ is rational, and hence V/T is rational as well.

Since $V \cong X \times D$, the morphism $\mathrm{pr}_A \colon V \to X$ has a rational section. Since $T \subset D$, the morphism $V/T \to X$ also has a rational section. Moreover, the variety X is a quotient of the variety V/T over the action of a two-dimensional torus $S = D/T$. Hence the variety V/T is birational to the product $S \times X$. The torus S is rational over k by Voskresenskii's theorem, having dimension 2. Thus the variety $X \times \mathbb{P}^2$ is also rational over k.

9.7. Further reading

To the best of our knowledge, up to now there have been no other examples of non-rational stably rational varieties except for those discussed in this chapter. In particular, by [**CT17**] every stably rational surface over a finite field is rational.

Part IV

The Hasse principle and its failure

CHAPTER 10

Minkowski–Hasse Theorem

In this chapter we describe a proof of the Minkowski–Hasse theorem. We will use several fundamental facts from class field theory, namely, exactness of the sequence (10.1). The proofs of these facts are too technical for our exposition; one can look them up, for instance, in [**CF67**, VII].

10.1. Preliminaries

Recall (see, for example, [**CF67**, II.1]) that a *valuation* on a field K is a homomorphism

$$K^* \to \mathbb{R}_{>0}, \quad x \mapsto |x|,$$

for which there is a positive constant $C \in \mathbb{R}_{>0}$ such that for every element $x \in K^*$, the inequality $|x| \leqslant 1$ implies the inequality $|1+x| \leqslant C$. Besides this, we always put $|0| = 0$ by definition. A valuation $|\cdot|_1$ is *equivalent* to a valuation $|\cdot|_2$ if there exists a positive constant $\lambda \in \mathbb{R}_{>0}$ such that $|\cdot|_1 = |\cdot|_2^{\lambda}$. A valuation is *non-archimedean* if one can take $C = 1$. This is equivalent to the fact that

$$|x+y| \leqslant \max\{|x|, |y|\}$$

for all $x, y \in K$. Otherwise, a valuation is called *archimedean*. It is easily seen that the equivalence relation on valuations preserves the property of being (non-)archimedean.

Note that a valuation $|\cdot|$ on a field K defines a metric on K by the formula

$$\rho(x,y) = |x-y|, \quad x, y \in K.$$

In turn, the metric ρ defines a topology on the field K such that addition, subtraction, multiplication, and division are continuous with respect to this topology.

EXERCISE 10.1.1. **Examples of valuations**

(i) Check that the absolute value $x \mapsto |x|$ defines an archimedean valuation on the field $\mathbb{Q}$. The completion of $\mathbb{Q}$ with respect to the corresponding metric is isomorphic to the field of real numbers $\mathbb{R}$.

(ii) Let p be a prime number. For every non-zero integer r put $\mathrm{ord}_p(r) = d$, where $r = p^d \cdot r'$ and r' is an integer coprime to p. Let $x \in \mathbb{Q}^*$ be represented as $x = m/n$. Put

$$\mathrm{ord}_p(x) = \mathrm{ord}_p(m) - \mathrm{ord}_p(n).$$

Prove that the function $x \mapsto p^{-\mathrm{ord}_p(x)}$ defines a non-archimedean valuation on the field $\mathbb{Q}$. The completion of $\mathbb{Q}$ with respect to the corresponding metric is isomorphic to the field of p-adic numbers $\mathbb{Q}_p$.

(iii) Show that for any discrete valuation

$$v\colon K^* \to \mathbb{Z}$$

(see §4.1) and a real number $\alpha > 1$, we have a well-defined non-archimedean valuation

$$|x|_v = \alpha^{-v(x)}, \quad x \in K^* .$$

In view of Exercise 10.1.1(iii), we denote a valuation on a field K by $|\cdot|_v$, even when the valuation is not induced by a discrete valuation, for example, if it is archimedean. Moreover, we usually abbreviate this notation to just v, still using the notation $|\cdot|_v$ for the actual function from K^* to $\mathbb{R}_{>0}$. Accordingly, the completion of K with respect to $|\cdot|_v$ (or, equivalently, with respect to v) is denoted by K_v.

Ostrowski's theorem (see, for example, [**CF67**, II.3]) states that all valuations of the field $\mathbb{Q}$ are given up to equivalence by the examples from Exercise 10.1.1.

A *global field* is either a finite extension of the field $\mathbb{Q}$ or a finite extension of the field $\mathbb{F}_p(T)$. There is a generalization of Ostrowski's theorem for an arbitrary global field (see, for example, [**CF67**, II.11, II.12]). In particular, it states that a global field has only finitely many archimedean valuations, and a global field of positive characteristic does not have them at all.

A *local field* is a completion K_v of a global field K with respect to a valuation v (cf. Definition 4.2.5). Thus, if v is archimedean, then the local field K_v is either the field of real numbers $\mathbb{R}$ or the field of complex numbers $\mathbb{C}$. If v is non-archimedean, then the local field K_v is a finite extension either of the field of p-adic numbers $\mathbb{Q}_p$, or of the field $\mathbb{F}_q((T))$, where q is a power of a prime number. Given an arbitrary finite collection S of valuations of a global field K, consider the diagonal embedding

$$K \to \prod_{v\in S} K_v .$$

The *weak approximation theorem* (see [**CF67**, II.6]) states that the image of K with respect to this embedding is dense (when $K = \mathbb{Q}$, this can be easily reduced to the Chinese remainder theorem).

Let Q be a smooth projective quadric over a global field K of characteristic different from 2. The *Minkowski–Hasse theorem* states that if for any valuation v of K the quadric Q has a point over the field K_v, then Q has a point over the field K. The goal of the next exercises is to deduce this theorem from certain results of class field theory.

10.2. Quadrics over local fields

EXERCISE 10.2.1. **Variation of coefficients**

Let F be a local field. One has a natural topology on F given by the valuation on F.

(o) Prove that the set of squares in F is open.

Hint. In the non-archimedean case, use either Hensel's lemma or the Taylor series of the function $\sqrt{1+t}$.

(i) Suppose that a smooth projective quadric $Q \subset \mathbb{P}^n$ over F is given by the equation

$$\sum_{i=0}^{n} a_i x_i^2 = 0 ,$$

where $a_i \in F^*$. Prove that there is a number $\varepsilon > 0$ that satisfies the following condition: if the inequalities $|b_i - a_i| < \varepsilon$ hold for all $0 \leqslant i \leqslant n$, then Q is isomorphic over F to the quadric Q' given by the equation

$$\sum_{i=0}^{n} b_i x_i^2 = 0 .$$

Hint. Using part (o), choose $\varepsilon > 0$ such that b_i/a_i is a square in F for all $0 \leqslant i \leqslant n$ whenever $|b_i - a_i| < \varepsilon$.

(ii) Let Q be a smooth projective quadric over F. Prove that Q remains isomorphic to itself under small deformations of coefficients of the equation of Q.

Hint. Prove that a matrix that gives a non-degenerate symmetric bilinear form can be transformed to a diagonal form in such a way that the coefficients of the diagonal matrix depend continuously on the coefficients of the initial matrix. Then use part (i).

EXERCISE 10.2.2. **Points on a quadric with smooth reductions**

Let $Q \subset \mathbb{P}^n_F$ be a positive-dimensional quadric over a non-archimedean local field F of characteristic different from 2 given by a polynomial with coefficients in the valuation ring $\mathcal{O}_F \subset F$. In addition, assume that the determinant of the corresponding matrix is an invertible element in $\mathcal{O}_F$. The goal of this exercise is to prove that $Q(F) \neq \varnothing$.

(o) Consider the reduction of the equation of the quadric Q modulo the maximal ideal $\mathfrak{m}_F \subset \mathcal{O}_F$. This defines a quadric $\overline{Q}$ over the finite residue field $k \cong \mathcal{O}_F/\mathfrak{m}_F$. Prove that the quadric $\overline{Q} \subset \mathbb{P}^n_k$ is smooth.

Hint. What is the determinant of the quadratic form over the field k that defines $\overline{Q}$?

(i) Prove that a smooth projective positive-dimensional quadric over a finite field has a point over it.

Hint. Either use the Chevalley–Warning theorem from Exercise 3.2.4, or prove this directly.

(ii) Choose a point

$$\bar{p} \in \overline{Q}(k) \subset \mathbb{P}^n_k .$$

Let $\bar{l}$ be a line that passes through the point $\bar{p}$ and is not contained in the embedded projective tangent space

$$\mathbb{T}_{\bar{p}}(\overline{Q}) \subset \mathbb{P}^n_k .$$

Lift arbitrarily the equation of the line $\bar{l}$ to the equation of a line l over $\mathcal{O}_F$. Prove that the quadric Q has a point that lies on the line l and is defined over F.

Hint. Use Hensel's lemma applied to the quadratic equation that defines the intersection of l and Q.

EXERCISE 10.2.3. **Points of quadrics over completions of a global field**
Let Q be a smooth projective positive-dimensional quadric over a global field K of characteristic different from 2. Prove that the quadric Q has a point over local fields K_v for almost all (that is, for all except a finite number of) valuations v of K.

Hint. Choose a homogeneous polynomial that defines the quadric Q in the corresponding projective space. Note that for almost all non-archimedean valuations v this defines a polynomial with coefficients in the valuation ring of K_v. Then use Exercise 10.2.2.

10.3. Reduction to the case $\dim(Q) = 1$

Let $Q \subset \mathbb{P}^n$ be a smooth projective quadric over a global field K of characteristic different from 2 such that $\dim(Q) \geqslant 3$, that is, $n \geqslant 4$. Consider a general projective subspace $\Lambda \subset \mathbb{P}^n$ of codimension 2 and the projection

$$\pi_\Lambda \colon \mathbb{P}^n \dashrightarrow \mathbb{P}^1$$

with center at Λ. Given a point $x \in \mathbb{P}^1$, by Q_x we denote the closure of the preimage $\pi_\Lambda^{-1}(x)$ in Q. Note that the quadric $Q \cap \Lambda$ is smooth because Λ is general. Clearly, $Q \cap \Lambda$ is contained in Q_x for every $x \in \mathbb{P}^1$.

EXERCISE 10.3.1. **Singular fibers of the morphism π_Λ**
Check that for all but two points $x \in \mathbb{P}^1$, the variety Q_x is a smooth quadric in $\mathbb{P}^{n-1}$.

Hint. Hyperplanes in $\mathbb{P}^n$ that contain Λ are parameterized by a line in the dual space $(\mathbb{P}^n)^\vee$, while tangent hyperplanes to Q are parameterized by a quadric in the dual space $(\mathbb{P}^n)^\vee$.

By the assumption of the Minkowski–Hasse theorem, for every valuation v of the field K there exists a point $p_v \in Q(K_v)$. Projection from p_v defines a birational equivalence $Q \dashrightarrow \mathbb{P}^{n-1}$ over the field K_v. Hence the set of K_v-points is Zariski dense in Q. Thus, by Exercise 10.3.1 we can assume that for every v, the point p_v does not lie on singular fibers of the map $\pi_\Lambda \colon Q \dashrightarrow \mathbb{P}^1$.

Let S be the set of all valuations v of the field K such that the quadric $Q \cap \Lambda$ does not have a K_v-point. Since the dimension of $Q \cap \Lambda$ is positive, by Exercise 10.2.3 applied to the quadric $Q \cap \Lambda$, the set S is finite.

EXERCISE 10.3.2. **Nice fibers of the map π_Λ**
For each valuation v of K, consider the point $x_v = \pi_\Lambda(p_v)$ in $\mathbb{P}^1(K_v)$. Prove that there exists a point $x \in \mathbb{P}^1(K)$ such that for every valuation $v \in S$, the quadric Q_x is isomorphic over the field K_v to the quadric Q_{x_v}.

Hint. Use the weak approximation theorem applied to the set of valuations S together with Exercise 10.2.1(ii).

EXERCISE 10.3.3. **Reduction to the case $\dim(Q) = 2$**
Show that in order to prove the Minkowski–Hasse theorem in the case $\dim(Q) \geqslant 3$, it is enough to prove it in the case $\dim(Q) = 2$.

Hint. In the above notation, we see that by Exercise 10.3.2 the fiber Q_x has a point over the field K_v for every $v \in S$. By construction, for any valuation $v \notin S$ the quadric $Q \cap \Lambda$ has a point over the field K_v, and so the quadric

$$Q_x \supset Q \cap \Lambda$$

has a point over K_v as well. Thus, Q_x has a point over K_v for each v. Since

$$\dim(Q_x) = \dim(Q) - 1 \geqslant 2\,,$$

induction on $\dim(Q)$ implies that the quadric Q_x has a point over the field K. Hence the quadric $Q \supset Q_x$ also has a point over the field K.

EXERCISE 10.3.4. **Reduction to the case** $\dim(Q) = 1$

Let $Q \subset \mathbb{P}^3$ be a smooth projective two-dimensional quadric over the global field K. Consider the field

$$L = K\big(\sqrt{\mathrm{d}(Q)}\big)\,,$$

where $\mathrm{d}(Q)$ is the discriminant of the quadric Q (see Definition 6.1.1). Let a conic C over the field L be a smooth plane section of the quadric Q_L. Prove that the Minkowski–Hasse theorem for the quadric Q over the field K follows from the Minkowski–Hasse theorem for the conic C over the field L.

Hint. For any valuation w of the field L, the completion L_w is an extension of the completion K_v, where v is defined as the restriction of w to K. Clearly, we have

$$L_w \cong K_v\big(\sqrt{\mathrm{d}(Q)}\big)\,.$$

Suppose that a quadric Q satisfies the conditions of the Minkowski–Hasse theorem, that is, Q has points over all fields K_v. By Exercise 6.1.4(ii), the conic C has points over all fields L_w. If the Minkowski–Hasse theorem holds for C, then C has a point over the field L. Using Exercise 6.1.4(ii) again, we obtain that the quadric Q has a point over the field K.

10.4. The case $\dim(Q) \leqslant 1$

Let K be a global field and take an element $\alpha \in \mathrm{Br}(K)$. One can show that for almost all valuations v, the image of α with respect to the natural map

$$\mathrm{Br}(K) \to \mathrm{Br}(K_v)$$

is equal to zero (cf. Exercise 4.4.3). One has the *exact sequence of class field theory* (see [**CF67**, VII.9.6, VII.11.2(bis)]):

$$0 \to \mathrm{Br}(K) \to \oplus_v \mathrm{Br}(K_v) \to \mathbb{Q}/\mathbb{Z} \to 0\,. \tag{10.1}$$

Recall that for non-archimedean valuations one has an isomorphism

$$\mathrm{res}\colon \mathrm{Br}(K_v) \xrightarrow{\sim} \mathbb{Q}/\mathbb{Z}$$

given by the residue map (see Exercise 4.2.6(i)), while for archimedean valuations one has

$$\mathrm{Br}(\mathbb{R}) \cong \tfrac{1}{2}\mathbb{Z}/\mathbb{Z}$$

(see Exercise 3.1.4), and $\mathrm{Br}(\mathbb{C}) = 0$. The second map in the sequence (10.1) corresponds to summation in the group $\mathbb{Q}/\mathbb{Z}$.

Let us assume that the characteristic of K is different from 2. By the correspondence between smooth conics and quaternion algebras (see Exercise 3.3.2(i)), exactness of the sequence (10.1) in the first term implies the Minkowski–Hasse theorem for the one-dimensional case. The fact that the composition of two maps vanishes on the 2-torsion subgroups in the case of $K = \mathbb{Q}$ is equivalent to the quadratic reciprocity law (see [**MP05**, 4.5.5]). Exactness in the middle term is an important result in class field theory and gives an explicit description of Brauer groups of global fields.

The case $\dim(Q) = 0$ is treated easily for the field $\mathbb{Q}$, while for an arbitrary global field K it is a non-trivial consequence of class field theory: an element in K^* is a square if and only if it is a square in K_v for every valuation v of the field K. For example, this assertion can be deduced from Chebotarev's density theorem (see [**CF67**, VIII]).

For the field $K = \mathbb{Q}$, one has a direct proof of the Minkowski–Hassse theorem in the one-dimensional case. Let us carry it out in terms of elements of the Brauer group. For short, let (a, b) denote the class in $\mathrm{Br}(K)$ of the quaternion algebra $A(a, b)$, where $a, b \in \mathbb{Q}^*$ (see Exercise 3.1.5).

Suppose that for every valuation v of the field $\mathbb{Q}$, the image of the class (a, b) with respect to the natural map

$$\mathrm{Br}(\mathbb{Q}) \to \mathrm{Br}(\mathbb{Q}_v)$$

is trivial. We want to prove that $(a, b) = 0$. Multiplying by squares and using bilinearity of the class (a, b) with respect to a and b (see Exercise 3.1.5(iii) and (iv)), we can assume that a and b are non-zero square-free integers and $|a| \leqslant |b|$. We will use induction on $M = |a| + |b|$.

EXERCISE 10.4.1. **Induction on M**

(o) Handle explicitly the case $M = 2$.

Hint. If either $a = 1$ or $b = 1$, then one can use Exercise 3.1.5(iii). To show that the case $a = b = -1$ is impossible, apply Exercise 3.1.5(ii) over the field $\mathbb{R}$.

(i) If $M > 2$, then $|b| > 1$. Consider an arbitrary prime divisor p of b. Prove that a is a square modulo p.

Hint. By Exercise 3.1.5(ii), the conic given by the equation

$$x^2 - ay^2 - bz^2 = 0$$

has a point over the field $\mathbb{Q}_p$.

Consequently, a is a square modulo b and we have an equality

$$a = c^2 - bb' , \tag{10.2}$$

where c and b' are integers. Moreover, changing the number b' if necessary, one can replace the number c in equation (10.2) by an integer in the interval between $-b/2$ and $b/2$. Thus we can assume that

$$|c| \leqslant \frac{|b|}{2} .$$

Deduce that $|b'| < |b|$.

(ii) Prove that if the number b' is chosen as described in part (i), then one has $(a, b) = (a, b')$.

Hint. Prove that

$$(a, b) + (a, b') = 0$$

using bilinearity of the class $(-, -)$ and also the Steinberg relation $(d, 1 - d) = 0$ for an arbitrary $d \neq 0, 1$; see Exercise 3.1.5(iii). Then use the equality $2(a, b) = 0$.

This completes the induction step.

Note that the above argument does not give a complete proof of the Minkowski–Hasse theorem even in the case of $K = \mathbb{Q}$, despite the fact that in Exercise 10.4.1 we provided an argument for conics over $\mathbb{Q}$. This is because, when making a reduction to conics, we need to pass to extensions of the field $\mathbb{Q}$ (see Exercise 10.3.4). There is a self-contained proof of the Minkowski–Hasse theorem over $\mathbb{Q}$ that does not use other number fields (see [**Ser70**, IV.3.2]). However, it uses Dirichlet's theorem on primes in arithmetic progressions, which is a subtle result and also has a direct relation to class field theory.

10.5. Other examples of the Hasse principle

Let $\mathcal{M}$ be a set of varieties over a fixed global field K (for instance, $\mathcal{M}$ may consist of all projective hypersurfaces of a given degree over K). One says that the *Hasse principle holds* for $\mathcal{M}$ if for every variety X in $\mathcal{M}$, the following is satisfied: the variety X has a point over the field K if and only if it has a point over the field K_v for any valuation v of K. For example, the Minkowski–Hasse theorem states that the Hasse principle holds for smooth quadrics (for singular quadrics it is also satisfied for trivial reasons).

EXERCISE 10.5.1. **Hasse principle for Severi–Brauer varieties**
Show that the Hasse principle holds for Severi–Brauer varieties (see Definition 2.2.6).

Hint. Use Exercise 3.3.4 and also the exact sequence of class field theory (10.1).

It follows from the exact sequence of class field theory (10.1) that for every finite Galois extension $K \subset L$, there is an exact sequence

$$0 \to \mathrm{Br}(L/K) \to \bigoplus_v \mathrm{Br}(L_w/K_v) \to \mathbb{Q}/\mathbb{Z}\,, \tag{10.3}$$

where for each valuation v of K one chooses an arbitrary valuation w of L that extends v from K to L (in fact, these two statements are equivalent).

EXERCISE 10.5.2. **The Hasse norm theorem**
Consider a finite cyclic Galois extension of global fields $K \subset L$ (that is, a Galois extension with a finite cyclic Galois group).

(i) Prove the *Hasse norm theorem*: an element $a \in K^*$ is a norm of an element $b \in L^*$ if and only if for every valuation v of the field K and for every valuation w of the field L that extends v, there is an element $b_w \in L_w^*$ such that

$$\mathrm{Nm}_{L_w/K_v}(b_w) = a\,.$$

Hint. Use the exact sequence (10.3) and the explicit form of cohomology groups for a cyclic group obtained in Exercise 1.1.5(i).

(ii) Prove that the Hasse principle holds for affine varieties in $R_{L/K}(\mathbb{G}_m)$ over K given by the equation

$$\mathrm{Nm}_{L/K}(x) = a, \quad x \in R_{L/K}(\mathbb{G}_m)\,,$$

where a is an element in K^*.

Hint. Use part (i).

10.6. Further reading

There are many other situations where the Hasse principle holds. Birch proved in [**Bir62**] that it holds for smooth complete intersections of r hypersurfaces in $\mathbb{P}^n$ over $\mathbb{Q}$ such that the degree of each hypersurface equals d, provided that one has

$$n \geqslant r(r+1)(d-1)2^{d-1}.$$

In particular, the Hasse principle holds for smooth intersections of two quadrics of dimension at least 10, and for smooth cubic hypersurfaces of dimension at least 15. There are a number of improvements of this result. It was proved in [**Dav63**] that an arbitrary (possibly singular) cubic hypersurface of dimension at least 15 has a rational point, whence it follows that the Hasse principle trivially holds. A similar result for smooth cubic hypersurfaces of dimension at least 9 was obtained in [**HB83**]. Later, the Hasse principle was established in [**Hoo13**] for smooth cubic hypersurfaces of dimension 8, though not all of them have rational points.

Intersection of two quadrics of lower dimension was treated in detail in [**CTSSD87a**] and [**CTSSD87a**], where the authors also prove the Hasse principle for some Châtelet surfaces.

The Hasse principle for families of varieties defined by norm forms arising from finite extensions of $\mathbb{Q}$ was proved for various cases in [**CTSD94**], [**HSW14**], [**BMS14**], and [**BM13**].

CHAPTER 11

Brauer–Manin Obstruction

In Chapter 10, we discussed the Hasse principle. In this chapter, we define the Brauer–Manin obstruction to the Hasse principle and, using it, construct a counterexample to the corresponding claim for curves of genus 1.

11.1. Definition of the Brauer–Manin obstruction

In what follows, we use Brauer groups of varieties and schemes (see Definition 4.4.1). For simplicity, we assume that K is a global field of characteristic zero, that is, K is a finite extension of $\mathbb{Q}$. Let X be a smooth projective variety over K. Then the Brauer group $\mathrm{Br}(X)$ of the variety X coincides with its unramified Brauer group (see Definitions 4.4.1 and 4.4.2, and also the discussion in §4.4). As above, given a valuation v of K, by K_v we denote the corresponding completion. For an element $\alpha \in \mathrm{Br}(X)$, a valuation v of K, and a point $p_v \in X(K_v)$, define the *local pairing*

$$\langle \alpha, p_v \rangle_v = \mathrm{res}\left(p_v^*(\alpha)\right) \in \mathbb{Q}/\mathbb{Z},$$

where the pull-back homomorphism

$$p_v^* \colon \mathrm{Br}(X) \to \mathrm{Br}(K_v)$$

is induced by the morphism $p_v \colon \mathrm{Spec}(K_v) \to X$ as described in §4.4. The map res for a non-archimedean valuation v is defined as in Exercise 4.2.6(i). If $K_v \cong \mathbb{R}$, then the map res is the natural embedding

$$\mathrm{Br}(K_v) \cong \mathbb{Z}/2\mathbb{Z} \hookrightarrow \mathbb{Q}/\mathbb{Z};$$

see Exercise 3.1.4(i).

Suppose that for any v, the variety X has a K_v-point. Define the *global pairing*

$$\langle -, - \rangle \colon \mathrm{Br}(X) \times \prod_v X(K_v) \to \mathbb{Q}/\mathbb{Z}, \quad \langle \alpha, (p_v) \rangle = \sum_v \langle \alpha, p_v \rangle_v .$$

In order to check that the latter pairing is well-defined, we need to show that the sum on the right-hand side is finite (cf. Exercise 4.4.3). Let $\mathcal{O}_K$ be the ring of integers in the field K. The field K is the fraction field of the ring $\mathcal{O}_K$, and $\mathcal{O}_K$ is a one-dimensional Noetherian integrally closed ring, that is, $\mathrm{Spec}(\mathcal{O}_K)$ is a regular one-dimensional scheme. Each non-zero prime ideal $\mathfrak{p}$ in $\mathcal{O}_K$ defines a discrete valuation on the field K. One can easily show that this correspondence is a bijection between the set of non-zero prime ideals $\mathfrak{p}$ in $\mathcal{O}_K$ and the set of equivalence classes of non-archimedean valuations v on K. Moreover, the valuation ring $\mathcal{O}_v$ of the complete field K_v is the completion of the local ring $(\mathcal{O}_K)_{\mathfrak{p}}$ with respect to its maximal ideal.

Furthermore, choose a projective model

$$\pi \colon \mathcal{X} \to \mathrm{Spec}(\mathcal{O}_K)$$

over $\mathcal{O}_K$ of the variety X (cf. §4.5). For any element $\alpha \in \mathrm{Br}(X)$, there is a non-empty open subset

$$U \subset \mathrm{Spec}(\mathcal{O}_K)$$

such that α is the restriction to the generic fiber of an element $\alpha_U \in \mathrm{Br}(\mathcal{X}_U)$, where

$$\mathcal{X}_U = \pi^{-1}(U)\,.$$

This follows directly from the definition of the Brauer group of a scheme in terms of Azumaya algebras; see Definition 4.4.1. Alternatively, one can use the description of the Brauer group of a scheme in terms of étale cohomology given in §A.6. Since the morphism π is projective, every point

$$p_v\colon \mathrm{Spec}(K_v) \to X$$

extends to a morphism

$$\tilde{p}_v\colon \mathrm{Spec}(\mathcal{O}_v) \to \mathcal{X}$$

by the valuative criterion of properness (see, e.g., [**Har77**, Theorem II.4.7]). Hence for all non-archimedean valuations v that correspond to points in U, the element $p_v^*(\alpha)$ equals the image of the element $\tilde{p}_v^*(\alpha_U)$ with respect to the map

$$\mathrm{Br}(\mathcal{O}_v) \to \mathrm{Br}(K_v)\,.$$

Therefore, for such v, we have $\langle \alpha, p_v\rangle_v = 0$ (in fact, it follows from Exercise 4.2.6(i) that $\mathrm{Br}(\mathcal{O}_v) = 0$).

EXERCISE 11.1.1. **Brauer–Manin obstruction**

(i) Show that the image of the natural map

$$\mathrm{Br}(K) \to \mathrm{Br}(X)$$

is contained in the kernel of the global pairing $\langle -,-\rangle$.

Hint. Use the exact sequence of class field theory (10.1).

Thus the global pairing factors through the quotient

$$\widetilde{\mathrm{Br}}(X) = \mathrm{Coker}\left(\mathrm{Br}(K) \to \mathrm{Br}(X)\right).$$

(ii) Prove that the image of the diagonal embedding

$$X(K) \to \prod_v X(K_v)$$

is contained in the kernel of the pairing $\langle -,-\rangle$.

Hint. Use again the exact sequence of class field theory (10.1).

(iii) Suppose that there is an element $\alpha \in \widetilde{\mathrm{Br}}(X)$ such that for any collection of local points

$$(p_v) \in \prod_v X(K_v)\,,$$

the global pairing $\langle \alpha, (p_v)\rangle$ is non-zero. Show that this implies that X does not have K-points.

In other words, there are embeddings of sets

$$X(K) \subset \mathrm{Br}(X)^{\perp} \subset \prod_v X(K_v)\,,$$

where $\mathrm{Br}(X)^{\perp}$ denotes the annihilator of the group $\mathrm{Br}(X)$ in $\prod_v X(K_v)$ with respect to the global pairing $\langle\cdot,\cdot\rangle$. Note that $\mathrm{Br}(X)^{\perp}$ is often called a *Brauer–Manin set*. Of course, if for some valuation v the variety X does not have K_v-points, or, equivalently, if $\prod_v X(K_v) = \varnothing$, then X does not have K-points either. The Brauer–Manin obstruction gives a stronger condition that implies emptiness of the set $X(K)$. Specifically, suppose that

$$\prod_v X(K_v) \neq \varnothing$$

but the set $\mathrm{Br}(X)^{\perp}$ is empty. Then X still does not have K-points. Emptiness of the set $\mathrm{Br}(X)^{\perp}$ is what is called the *Brauer–Manin obstruction* to the existence of K-points on X. This obstruction was introduced in [**Man71**].

11.2. Computation of the Brauer–Manin obstruction

Our next goal is to describe the group $\widetilde{\mathrm{Br}}(X)$ defined in Exercise 11.1.1 and to develop a way to construct its non-trivial elements in some interesting cases. We use the following notation: given a group of the form $\mathrm{Br}(-)$ with a natural map

$$b\colon \mathrm{Br}(K) \to \mathrm{Br}(-),$$

denote the quotient $\mathrm{Br}(-)/b\big(\mathrm{Br}(K)\big)$ by $\widetilde{\mathrm{Br}}(-)$. In what follows, we assume that the variety X is geometrically irreducible (that is, $X_{\bar{K}}$ is irreducible). Given a field extension $K \subset F$, we denote the field of rational functions on X_F by $F(X)$. We will use the following non-trivial fact from class field theory: one has

$$H^3\big(G_K, \bar{K}^*\big) = 0$$

for every global field K (see [**CF67**, VII.11.4]).

EXERCISE 11.2.1. **The group $\widetilde{\mathrm{Br}}(X)$**

(i) Construct a canonical isomorphism

$$\widetilde{\mathrm{Br}}\big(\bar{K}(X)/K(X)\big) \cong H^2\big(G_K, \bar{K}(X)^*/\bar{K}^*\big).$$

Hint. Consider an exact sequence of G_K-modules

$$1 \to \bar{K}^* \to \bar{K}(X)^* \to \bar{K}(X)^*/\bar{K}^* \to 1;$$

cf. Exercise 3.3.5(i). Use the vanishing $H^3(G_K, \bar{K}^*) = 0$.

(ii) Show that there is an exact sequence

(11.1)
$$0 \to H^1\big(G_K, \mathrm{Pic}(X_{\bar{K}})\big) \to \widetilde{\mathrm{Br}}\big(\bar{K}(X)/K(X)\big) \overset{\oplus \mathrm{res}_D}{\longrightarrow} \bigoplus_{D \subset X} \mathrm{Hom}\,\big(G_{K(D)}, \mathbb{Q}/\mathbb{Z}\big),$$

where the direct sum is taken over all reduced irreducible divisors D on X.

Hint. Consider an exact sequence of G_K-modules

$$1 \to \bar{K}(X)^*/\bar{K}^* \to \mathrm{Div}(X_{\bar{K}}) \to \mathrm{Pic}(X_{\bar{K}}) \to 0;$$

cf. Exercise 3.3.5(ii). Then use the fact that the embedding of fields

$$K_D \subset K(D)$$

induces a surjective map between Galois groups

$$G_{K(D)} \to G_{K_D}$$

and an embedding

$$\mathrm{Hom}\,\big(G_{K_D}, \mathbb{Q}/\mathbb{Z}\big) \hookrightarrow \mathrm{Hom}\,\big(G_{K(D)}, \mathbb{Q}/\mathbb{Z}\big).$$

Finally, apply part (i) and Exercise 2.5.1(iii).

(iii) Show that there is an exact sequence

$$0 \to H^1\big(G_K, \mathrm{Pic}(X_{\bar{K}})\big) \to \widetilde{\mathrm{Br}}(X) \to \mathrm{Br}(X_{\bar{K}})\,. \tag{11.2}$$

Hint. It follows from Definition 4.4.2 and the discussion after it that there is an exact sequence

$$0 \to \widetilde{\mathrm{Br}}(X) \to \widetilde{\mathrm{Br}}\big(K(X)\big) \stackrel{\oplus \mathrm{res}_D}{\longrightarrow} \bigoplus_{D \subset X} \mathrm{Hom}\left(G_{K(D)}, \mathbb{Q}/\mathbb{Z}\right). \tag{11.3}$$

Part (ii) implies that there is an embedding

$$H^1\big(G_K, \mathrm{Pic}(X_{\bar{K}})\big) \hookrightarrow \widetilde{\mathrm{Br}}\big(\bar{K}(X)/K(X)\big)\,. \tag{11.4}$$

It follows from the exact sequences (11.1) and (11.3) that the image of the group $H^1\big(G_K, \mathrm{Pic}(X_{\bar{K}})\big)$ under the composition of (11.4) with the natural embedding $\widetilde{\mathrm{Br}}\big(\bar{K}(X)/K(X)\big) \hookrightarrow \widetilde{\mathrm{Br}}\big(K(X)\big)$ is contained in the subgroup

$$\widetilde{\mathrm{Br}}(X) \subset \widetilde{\mathrm{Br}}\big(K(X)\big)\,.$$

Besides this, we have a commutative diagram

$$\begin{array}{ccc} \widetilde{\mathrm{Br}}(X) & \hookrightarrow & \widetilde{\mathrm{Br}}\big(K(X)\big) \\ \downarrow & & \downarrow \\ \mathrm{Br}\left(X_{\bar{K}}\right) & \hookrightarrow & \mathrm{Br}\left(\bar{K}(X)\right) \end{array}$$

This implies exactness of the sequence (11.2) in the middle term.

(iv) Suppose that $\mathrm{Br}(X_{\bar{K}}) = 0$. (For example, $X_{\bar{K}}$ is a smooth projective curve or a rational smooth projective variety over $\bar{K}$. Another example of such a situation arises when we have $H^2(X, \mathcal{O}_X) = 0$, and for the scalar extension $X_{\mathbb{C}}$ of the variety X with respect to some (equivalently, any) embedding $K \hookrightarrow \mathbb{C}$ one has

$$H^3\big(X(\mathbb{C}), \mathbb{Z}\big)_{tors} = 0\,;$$

see §A.7.) Show that in this case, there is an isomorphism

$$H^1\big(G_K, \mathrm{Pic}(X_{\bar{K}})\big) \cong \widetilde{\mathrm{Br}}(X)\,.$$

Hint. This follows immediately from part (iii).

(v) Suppose that $\mathrm{Br}(X_{\bar{K}}) = 0$ and the group $\mathrm{Pic}(X_{\bar{K}})$ is torsion-free (in particular, the group of algebraically trivial divisors $\mathrm{Pic}^0(X_{\bar{K}})$ vanishes). Show that the group $\widetilde{\mathrm{Br}}(X)$ is finite.

Hint. Use the fact that in this case the Picard group $\mathrm{Pic}(X_{\bar{K}})$ is isomorphic to the Neron–Severi group of $X_{\bar{K}}$, and hence it is finitely generated. Thus, $\mathrm{Pic}(X_{\bar{K}})$ is a free finitely generated abelian group. Now use Exercise 2.3.9(iii).

(vi) Suppose that $\mathrm{Br}(X_{\bar{K}}) = 0$ and $\mathrm{Pic}(X_{\bar{K}}) \cong \mathbb{Z}$. (For example, X is a smooth quadric of dimension at least 3, or a Severi–Brauer variety.) Show that

$$\widetilde{\mathrm{Br}}(X) = 0\,.$$

Hint. Use part (iv) and Exercise 2.3.7(iii).

Note that the exact sequence of Exercise 11.2.1(iii) can also be deduced from the description of the Brauer group of a scheme in terms of étale cohomology (see §A.6) and the Hochschild–Serre spectral sequence for étale cohomology,

$$H^i\big(G_K, H^j_{\acute{e}t}(X_{\bar{K}}, \mathbb{G}_m)\big) \implies H^{i+j}_{\acute{e}t}(X, \mathbb{G}_m)\,.$$

EXERCISE 11.2.2. **Element in the Brauer group constructed from a divisor**

Let $K \subset L$ be a finite Galois extension with a cyclic Galois group G of order n. Fix a generator s of the group G.

(o) Let $D \in \mathrm{Div}(X_L)$ be a divisor, and let $f \in K(X)^*$ be a rational function such that

$$\sum_{g\in G} g(D) = (f) \in \mathrm{Div}(X)\,,$$

where (f) denotes the divisor of the function f. Prove that D defines a class

$$[D] \in H^1\big(G, \mathrm{Pic}(X_L)\big)\,.$$

Hint. Use Exercise 1.1.5(i).

(i) Check that $K(X) \subset L(X)$ is a Galois extension and that its Galois group is canonically isomorphic to G. Let $A(f)$ denote the cyclic algebra over the field $K(X)$ that corresponds to the element $f \in K(X)^*$, the extension $K(X) \subset L(X)$, and the generator s of the Galois group G (see Exercise 3.1.6(i)). Prove that the image of the class $[D]$ in the group $\widetilde{\mathrm{Br}}\big(K(X)\big)$ with respect to the composition

$$H^1\big(G, \mathrm{Pic}(X_L)\big) \to H^1\big(G_K, \mathrm{Pic}(X_{\bar{K}})\big) \to \widetilde{\mathrm{Br}}(X) \to \widetilde{\mathrm{Br}}\big(K(X)\big)$$

is equal to the class $[A(f)]$.

Hint. Use the construction of the isomorphism in Exercise 11.2.1(ii) together with Exercise 1.1.6(iv).

In particular, the class $[A(f)]$ belongs to the subgroup

$$\mathrm{Br}(X) = \mathrm{Br}^{nr}\left(K(X)\right) \subset \mathrm{Br}\left(K(X)\right).$$

(ii) Show directly that the class $[A(f)]$ belongs to the subgroup

$$\mathrm{Br}^{nr}\left(K(X)\right) \subset \mathrm{Br}\left(K(X)\right);$$

that is, check explicitly that the residue of the class $[A(f)] \in \mathrm{Br}\left(K(X)\right)$ is zero for all divisors on X.

Hint. By Exercise 4.2.3(ii), the equality

$$\mathrm{res}_E\left([A(f)]\right) = 0$$

holds for any reduced irreducible divisor E on X which is not contained in the support $\mathrm{Supp}(f)$ of the principal divisor (f). Given an irreducible component D of $\mathrm{Supp}(f)$, consider a rational function $h \in L(X)^*$ such that

$$(h) = D - D'\,,$$

where a (not necessarily effective) divisor D' does not have common components with the divisor

$$(f) = \sum_{g \in G} g(D).$$

Then use the fact that the class of the cyclic algebra $A\big(\mathrm{Nm}(h)\big)$ equals zero in the group $\mathrm{Br}\,\big(K(X)\big)$; see Exercise 3.1.6(o). Here Nm denotes the Galois norm for the field extension $K(X) \subset L(X)$. Also use the equality

$$[A(f)] = \big[A\big(\mathrm{Nm}(h)\big)\big] + \big[A\big(f \cdot \mathrm{Nm}(h)^{-1}\big)\big]$$

in the group $\mathrm{Br}\,\big(K(X)\big)$.

In order to compute the Brauer–Manin obstruction, it would be useful, given an element $\alpha \in \mathrm{Br}(X)$, to find conditions on a valuation v of K that imply that the local pairing $\langle \alpha, p_v \rangle_v$ equals zero for any point $p_v \in X(K_v)$. We do this in the case $\alpha = [A(f)]$, where f and $A(f)$ are as in Exercise 11.2.2. Note that the pairing $\langle \alpha, p_v \rangle_v$ is defined for the element $\alpha \in \mathrm{Br}(X)$ and not for its image in the quotient $\widetilde{\mathrm{Br}}(X)$. Nevertheless, the property of *being contained in the kernel of the global pairing* depends only on the image of an element of the group $\mathrm{Br}(X)$ in the quotient; see Exercise 11.1.1(i).

Let us introduce more notation. Let $\mathcal{X}$ be a projective model of the variety X over $\mathcal{O}_K$ (see §4.5 and §11.1). Given a non-archimedean valuation v, define a finite field $\kappa_v = \mathcal{O}_K/\mathfrak{p}$, where $\mathfrak{p}$ is a prime ideal in $\mathcal{O}_K$ that corresponds to the valuation v. Let $\mathcal{X}_v$ denote the product

$$\mathrm{Spec}(\mathcal{O}_v) \times_{\mathrm{Spec}(\mathcal{O}_K)} \mathcal{X}$$

and let $\mathcal{X}(v)$ denote the fiber of the scheme $\mathcal{X}_v$ at the closed point

$$\mathrm{Spec}(\kappa_v) \hookrightarrow \mathrm{Spec}(\mathcal{O}_v).$$

Note that $\mathcal{X}(v)$ is also the fiber of the scheme $\mathcal{X}$ at the closed point

$$\mathrm{Spec}(\kappa_v) \hookrightarrow \mathrm{Spec}(\mathcal{O}_K).$$

Given a rational function f in $K(X)^*$ (respectively, in $K_v(X)^*$), we denote by $\mathcal{S}upp(f)$ the support of the divisor of the function f on the scheme $\mathcal{X}$ (respectively, on $\mathcal{X}_v$).

Let $K \subset L$ be an extension of global fields, and let v be a valuation on K. There is a decomposition

$$L \otimes_K K_v \cong \prod_{i=1}^{r} L_{w_i} \tag{11.5}$$

into a product of r local fields (note that all assertions from Exercise 4.1.2 hold for arbitrary valuations, not just discrete ones). We say that a valuation v *splits in* L if all fields L_{w_i} are isomorphic to K_v. A non-archimedean valuation v is *unramified in* L if for all i the extension of local fields $K_v \subset L_{w_i}$ is unramified (see §4.1). Note that all valuations of K except a finite number are unramified in L (see [**Lan64**, Proposition III.2.8]).

EXERCISE 11.2.3. **Triviality of the local pairing**

We use the notation and assumptions from Exercise 11.2.2. As above, v is a valuation on K.

(i) Show that the image of the class $[A(f)]$ with respect to the natural map

$$\mathrm{Br}\left(K(X)\right) \to \mathrm{Br}\left(K_v(X)\right)$$

is the class of the cyclic algebra over $K_v(X)$ that corresponds to the extension $K_v(X) \subset L_w(X)$, the element

$$f \in K(X)^* \subset K_v(X)^* ,$$

and the generator s^r of the subgroup

$$\mathrm{Gal}(L_w/K_v) \subset \mathrm{Gal}(L/K)$$

(see Exercise 4.1.2(v)), where w is any of the valuations w_i, $1 \leqslant i \leqslant r$, that appear in (11.5).

Hint. Use Exercises 1.2.2(iv) and 3.1.6(o).

(ii) Suppose that the valuation v splits in L. Show that in this case, the class of the algebra $A(f)$ in $\mathrm{Br}\left(K_v(X)\right)$ equals zero.

Hint. Use part (i).

(iii) Suppose that the function f is regular at a point $p_v \in X(K_v)$ and that $f(p_v) \neq 0$. Show that

$$p_v^*\left([A(f)]\right) \in \mathrm{Br}(K_v)$$

is the cyclic algebra that corresponds to the extension $K_v \subset L_w$, the element $f(p_v) \in K_v^*$, and the generator s^r of the group $\mathrm{Gal}(L_w/K_v)$, where r and w are as in part (i).

Hint. Use part (i), the explicit description of a cyclic algebra in Exercise 3.1.6(i), and the definition of the pull-back homomorphism p_v^* given at the end of Chapter 4.

(iv) Suppose that v is a non-archimedean valuation and v is unramified in L. As we saw above in §11.1, any point $p_v \in X(K_v)$ defines a morphism

$$p_v \colon \mathrm{Spec}(K_v) \to \mathcal{X}_v ,$$

which extends to a morphism

$$\tilde{p}_v \colon \mathrm{Spec}(\mathcal{O}_v) \to \mathcal{X}_v .$$

Suppose that for a point p_v, the support $\mathcal{S}upp(f)$ does not intersect the image $\mathrm{Im}(\tilde{p}_v)$ on $\mathcal{X}_v$. Prove that

$$\langle [A(f)], p_v \rangle_v = 0 .$$

Hint. First show that the condition on p_v implies that the function f is regular at p_v and we have

$$f(p_v) \in \mathcal{O}_v^* \subset K_v .$$

Then use part (iii) and Exercise 4.2.3(ii).

(v) Suppose that a non-archimedean valuation v is unramified in L and the support $\mathcal{S}upp(f)$ does not contain the fiber $\mathcal{X}(v) \subset \mathcal{X}_v$. Prove that for every point $p_v \in X(K_v)$, there is a rational function $h \in L_w(X)^*$ such that the support $\mathcal{S}upp\left(f \cdot \mathrm{Nm}(h)^{-1}\right)$ does not intersect the image $\mathrm{Im}(\tilde{p}_v)$ on $\mathcal{X}_v$, where Nm denotes the Galois norm for the field extension $K_v(X) \subset L_w(X)$.

Hint. Put
$$\mathcal{X}_w = \mathrm{Spec}(\mathcal{O}_w) \times_{\mathrm{Spec}(\mathcal{O}_v)} \mathcal{X}\,.$$
Consider the divisor $\mathcal{D}$ on the scheme $\mathcal{X}_w$ defined as the closure of the divisor $D \subset X_L$ from Exercise 11.2.2(o). Apply the method from the solution of Exercise 11.2.2(ii) to the divisor $\mathcal{D}$ on $\mathcal{X}_w$ using the relative projective embedding of the scheme $\mathcal{X}$ over $\mathrm{Spec}(\mathcal{O}_K)$.

(vi) Let S be the set that consists of all real valuations of K that do not split in L and all non-archimedean valuations v of K such that either v is unramified in L or the support $\mathcal{S}upp(f)$ contains the fiber $\mathcal{X}(v)$. Show that the set S is finite. Prove that if $v \notin S$, then for every point $p_v \in X(K_v)$ there is an equality
$$\langle [A(f)], p_v\rangle_v = 0\,.$$
Hint. For non-archimedean valuations use parts (iv) and (v), and also the equality
$$[A(f \cdot \mathrm{Nm}(h)^{-1})] = [A(f)]$$
in the group $\mathrm{Br}\left(K_v(X)\right)$; cf. the hint for Exercise 11.2.2(ii). For archimedean valuations use part (ii).

Note that if the ring $\mathcal{O}_K$ is factorial (e.g., if $K = \mathbb{Q}$), then there is an element $c \in K^*$ such that the support $\mathcal{S}upp(c \cdot f)$ does not contain the fiber $\mathcal{X}(v)$ for any non-archimedean valuation v.

11.3. Brauer–Manin obstruction for a genus-one curve

In this section, we use the Brauer–Manin obstruction to construct a counterexample to the Hasse principle for curves of genus one. Recall that an *elliptic curve* over a field F is a curve of genus one over F with a marked F-point.

EXERCISE 11.3.1. **Curves of genus one**

(i) Show that any smooth projective curve E of genus one is a torsor (see Definition 2.6.1) under an elliptic curve.

Hint. Consider the Jacobian $J(E)$ of the curve E and use the Riemann–Roch theorem to define the action of $J(E)$ on E.

(ii) Show that for any elliptic curve E over a finite field $\mathbb{F}_q$, there is an equality
$$H^1\left(G_{\mathbb{F}_q}, E(\bar{\mathbb{F}}_q)\right) = 0\,.$$
Hint. First prove surjectivity of the morphism
$$\Phi_q - \mathrm{id}\colon E_{\bar{\mathbb{F}}_q} \to E_{\bar{\mathbb{F}}_q}$$
by taking its differential, where Φ_q denotes the $\bar{\mathbb{F}}_q$-linear Frobenius morphism, defined by raising coordinates of points to the power q. Then use Exercise 2.3.8.

(iii) Prove that every smooth projective curve of genus one over a finite field $\mathbb{F}_q$ has an $\mathbb{F}_q$-point.

Hint. Use parts (i) and (ii), and also Exercise 2.6.2(i).

(iv) Prove that any smooth projective plane curve of genus one with smooth reduction (modulo the maximal ideal in the valuation ring) over a non-archimedean local field has a point over this field.

Hint. Use part (iii) and Hensel's lemma as in Exercise 10.2.2(ii).

In fact, an analogous statement holds for any (not necessarily plane) smooth projective curve with smooth reduction.

Note that Lang proved surjectivity of the morphism

$$g \mapsto \Phi_q(g) \cdot g^{-1}$$

for any connected algebraic group over the field $\bar{\mathbb{F}}_q$ (see [**Lan56**] and also a shorter Steinberg's proof in [**Ste77**]).

Let X be a smooth projective curve of genus one over a global field K of characteristic zero (we are not assuming that X has a K-point). Suppose that X has an effective reduced divisor $F \in \mathrm{Div}(X)$ of degree two, that is, there is a morphism $\psi\colon X \to \mathbb{P}^1$ over K of degree two (the morphism ψ is given by the linear system $|F|$). Denote by $Z \in \mathrm{Div}(X)$ the ramification divisor of the morphism ψ. Suppose that Z is irreducible over K. Suppose also that there exists a quadratic extension $K \subset L$ such that

$$Z = E + \sigma(E)\,,$$

where $E \in \mathrm{Div}(X_L)$ is an effective reduced divisor of degree two and σ denotes the non-trivial involution of the field L over K.

EXERCISE 11.3.2. **Construction of an element in the Brauer group**

(i) Show that there is a rational function $f \in K(X)^*$ such that

$$Z - 2F = (f)\,,$$

where, as above, (f) denotes the divisor of the function f. (Note that we do not use the assumption on the existence of the field L here.)

(ii) Let $a \in K$ be an element such that $L = K(\sqrt{a})$. Prove that the class

$$\alpha \in \mathrm{Br}\left(K(X)\right)$$

of the quaternion algebra $A_{-1}(a, f)$ belongs to the subgroup

$$\mathrm{Br}(X) \subset \mathrm{Br}\left(K(X)\right).$$

Hint. Use Exercise 11.2.2 with $D = E - F$.

Now we construct explicitly a curve of genus one that gives a counterexample to the Hasse principle. (This example was first found by C.-E. Lind in [**Lin40**] and by H. Reichardt in [**Rei42**].) Consider a smooth projective curve X over $\mathbb{Q}$ such that its open affine subset U is given in $\mathbb{A}^2$ with coordinates x and y by the equation

$$2y^2 = x^4 - 17\,.$$

Note that X is a union of the affine chart U and another affine chart V, which is naturally identified with a curve given in $\mathbb{A}^2$ with coordinates z and w by the equation

$$2z^2 = 1 - 17w^4\,.$$

One has an effective reduced divisor $F = \{w = 0\}$ on X. In the notation introduced before Exercise 11.3.2, we have

$$Z = \{y = 0\} = \{z = 0\} \subset U \cap V, \quad L = \mathbb{Q}(\sqrt{17}), \quad f = y = \frac{z}{w^2}.$$

Let v_∞ denote the (unique) archimedean valuation of $\mathbb{Q}$.

EXERCISE 11.3.3. **Brauer–Manin obstruction for the curve X**

(o) Show that $p = 17$ is the only prime number p such that the valuation v_p of $\mathbb{Q}$ is ramified in the extension $\mathbb{Q} \subset \mathbb{Q}(\sqrt{17})$.

Hint. We are interested in the extension of local fields

$$\mathbb{Q}_p \subset \mathbb{Q}_p(\sqrt{17})\,.$$

First check that for $p \notin \{2, 17\}$, this extension is unramified (see §4.1). Then check that the valuation v_2 splits in the extension $\mathbb{Q} \subset \mathbb{Q}(\sqrt{17})$ by using the fact that 17 is a square in the field $\mathbb{Q}_2$.

(i) Show that for every valuation

$$v \notin \{v_\infty, v_2, v_{17}\}$$

of $\mathbb{Q}$, there is a point $p_v \in X(\mathbb{Q}_v)$.

Hint. Use Exercise 11.3.1(iv).

(ii) Show that for every valuation

$$v \in \{v_\infty, v_2, v_{17}\}\,,$$

there is a point $p_v \in X(\mathbb{Q}_v)$.

Hint. For $v = v_\infty$, the assertion is obvious. For $v = v_2$, we have $Z(\mathbb{Q}_2) \neq \varnothing$, because 17 is a fourth power in $\mathbb{Q}_2$. For $v = v_{17}$, we have $F(\mathbb{Q}_{17}) \neq \varnothing$, because 2 is a square in $\mathbb{Q}_{17}$.

(iii) Show that for every valuation $v \neq v_{17}$ and every point $p_v \in X(\mathbb{Q}_v)$, one has

$$\langle \alpha, p_v \rangle_v = 0\,,$$

where α is defined as in Exercise 11.3.2(ii).

Hint. Use part (o) and Exercise 11.2.3(vi).

(iv) Show that for every point $p_{17} \in X(\mathbb{Q}_{17})$, one has

$$\langle \alpha, p_{17} \rangle_{v_{17}} \neq 0\,.$$

Hint. A point $p_{17} \in X(\mathbb{Q}_{17})$ has coordinates in $\mathbb{Z}_{17}$ either on U or on V. For $p_{17} \in U(\mathbb{Z}_{17})$, the local pairing

$$\langle \alpha, p_{17} \rangle_{v_{17}} \in \mathbb{Q}/\mathbb{Z}$$

corresponds to the Hilbert symbol $\big(17, y(p_{17})\big)_{17}$ as described in Exercise 4.2.3(iv). In order to compute the Hilbert symbol, show that

$$v_{17}\big(y(p_{17})\big) = 0\,,$$

while $y(p_{17})$ is not a square modulo 17, because 2 is not a fourth power in $\mathbb{F}_{17}$. A point $p_{17} \in V(\mathbb{Z}_{17})$ is considered similarly.

(v) Prove that $X(\mathbb{Q}) = \varnothing$. Thus, X provides a counterexample to the Hasse principle.

Hint. Use parts (iii) and (iv), and also Exercise 11.1.1(iii).

11.4. Further reading

Most of the examples below concern varieties defined over the field of rational numbers.

The Hasse principle fails in many interesting cases. Its failure was shown in [**Sel51**] for an explicit curve of genus one different from the curve considered in §11.3. It was proved in [**Isk71b**] that the Hasse principle fails for intersections of two quadrics in $\mathbb{P}^4$. Also, this was shown for some diagonal smooth cubic threefolds in [**CG66**]. In all these examples one checks for the absence of rational points by applying the Brauer–Manin obstruction, namely, by showing that the Brauer–Manin set is empty.

There are families of varieties for which the Brauer–Manin obstruction is the only obstruction to the existence of rational points; that is, a variety from such a family has a rational point if and only if its Brauer–Manin set is non-empty. Examples were constructed among homogeneous spaces under algebraic groups in [**San81**] and [**Bor96**], and for smooth intersection of two quadrics and Châtelet surfaces in [**CTSSD87b**], [**CTSSD87c**], and [**SS91**]. In all these examples it was actually proved that the set of rational points is dense in the Brauer–Manin set with respect to the product topology. A conjecture of Colliot-Thélène predicts that this should hold for all smooth projective geometrically rationally connected varieties; see [**CT03**].

Another particular case of this conjecture concerns fibrations in geometrically rationally connected varieties over a projective line. Significant results towards this case were obtained in [**Har94**], [**BM13**], and [**HW16**].

However, there are examples where the Brauer–Manin set is not empty but still there are no rational points. The first example was constructed by Skorobogatov in [**Sko99**] and is provided by a bielliptic surface. A generalization was given later in [**HS02**]. In these examples the absence of rational points was shown with the help of a descent argument. In [**Poo10**] an étale Brauer–Manin set was introduced; this is a certain subset of the Brauer–Manin set which also contains the set of all rational points. The counterexample from [**Sko99**] is naturally interpreted in terms of the étale Brauer–Manin set. It is known from the papers [**Bor96**], [**Har02**], [**Dem09**], [**Sko09**], and [**Cao17**] that the corresponding étale Brauer–Manin obstruction is equivalent to the obstruction from descent, which, in turn, is equivalent to its own iterated version.

Furthermore, there are examples of varieties containing non-trivial étale Brauer–Manin set and without rational points: a threefold in [**Poo10**], a surface of general type in [**HS14**], a conic bundle over a curve in [**CTPS16**], and a Beauville surface in [**Sme17**].

There are also papers containing explicit descriptions of the Brauer–Manin set in concrete examples; let us just mention [**KT11**].

One can find thorough surveys on Brauer–Manin obstruction and related questions in [**Sko01**], [**Poo17**], and [**Wit16**].

APPENDIX A

Étale Cohomology

In this appendix, we provide some general notions and statements from the theory of étale cohomology that are related to Brauer groups of varieties. An excellent survey of étale cohomology is given in [**Dan96**], and a detailed exposition of all the facts we need together with their proofs is contained in [**Mil80**]. Below we give explicit references to both sources. We also recommend the book [**FK88**] for further reading.

A.1. Étale coverings

An *étale morphism* between schemes is a smooth morphism of relative dimension zero. There are many equivalent definitions of an étale morphism; see, for example, [**Dan96**, 4.2.1] or [**Mil80**, I.3]. An extension of fields corresponds to an étale morphism between their spectra if and only if it is a finite separable extension. An *étale covering* of a scheme X is a collection of étale morphisms $\{U_i \to X\}$ such that the union of their images equals X; see [**Mil80**, II.1]. If X is quasi-compact, then usually it is enough to consider only finite étale coverings, that is, finite collections $\{U_i \to X\}$. In this case, it is more convenient to pass to the disjoint union $U = \coprod_i U_i$ and the surjective étale morphism $U \to X$. The class of étale coverings is closed under taking compositions and fibered products, that is, étale coverings form a Grothendieck topology; see [**Mil80**, Remark II.1.1].

One of the main reasons why the numerous successful applications of étale topology are possible is the following lemma due to M. Artin: for any point x on a smooth complex algebraic variety, there exists a neighborhood U of x in Zariski topology such that all its homotopy groups starting from the second one are trivial; see [**Dan96**, 3.4.3]. Thus the only obstruction to U being contractible is its fundamental group. From the "profinite" point of view one gets rid of this obstruction by taking various finite unramified covers of U, that is, by taking étale neighborhoods of x.

A.2. Sheaves in the étale topology

An *étale presheaf of sets* $\mathcal{F}$ on a scheme X is a contravariant functor $U \mapsto \mathcal{F}(U)$ from the category of étale schemes over X, that is, étale morphisms $U \to X$, to the category of sets ([**Dan96**, 4.3.1], [**Mil80**, II.1]). An étale presheaf of sets is an *étale sheaf* if it sends direct limits to inverse limits; see [**Dan96**, 4.3.2]. One obtains an equivalent definition requiring that for any étale covering $\{V_i \to U\}$ of an étale scheme U over X, the map

$$\mathcal{F}(V) \to \prod_i \mathcal{F}(V_i)$$

is the equalizer of two maps from $\coprod_i \mathcal{F}(V_i)$ to $\coprod_{i,j} \mathcal{F}(V_i \times_U V_j)$ that correspond to projections from $V_i \times_U V_j$ to V_i and V_j; see [**Mil80**, II.1]. For example, the category of étale sheaves on the spectrum of a field K is equivalent to the category of discrete sets with a continuous action of the Galois group G_K of K; see [**Dan96**, 4.3.2].

An arbitrary scheme S represents an étale sheaf (on any scheme X)

$$U \mapsto h_S(U) = \mathrm{Hom}(U, S)$$

([**Dan96**, Example 4.3.1.1], [**Mil80**, Corollary II.1.7]). In particular, for each group scheme, one defines an étale sheaf of groups, which is usually denoted in the same way as the initial group scheme. For example, one has abelian étale sheaves of groups $\mathbb{G}_m$ and μ_l represented by commutative group schemes $\mathrm{Spec}(\mathbb{Z}[T, T^{-1}])$ and $\mathrm{Spec}\left(\mathbb{Z}[T]/(T^l - 1)\right)$, respectively ([**Dan96**, Example 4.4.1.2], [**Mil80**, Example II.2.18(b)]). For any smooth morphism of schemes $S \to S'$, the morphism of étale sheaves $h_S \to h_{S'}$ is surjective; see [**Dan96**, 4.3.3].

An arbitrary quasi-coherent sheaf $\mathcal{F}$ on X defines an étale sheaf $\mathcal{F}_{\acute{e}t}$ on X by the formula $\mathcal{F}_{\acute{e}t}(U) = \Gamma(U, \pi^*\mathcal{F})$, where $\pi\colon U \to X$ is an étale morphism ([**Dan96**, Example 4.3.1.2], [**Mil80**, Corollary II.1.6]).

A.3. Cohomology of étale sheaves of abelian groups

Cohomology groups of abelian sheaves on a scheme X are right derived functors of the left exact functor of global sections $\mathcal{F} \mapsto \mathcal{F}(X)$ ([**Dan96**, 4.4.2], [**Mil80**, Definition III.1.5(a)]). Usually, one denotes the cohomology of an abelian étale sheaf $\mathcal{F}$ by $H^i_{\acute{e}t}(X, \mathcal{F})$. One can also consider Čech cohomology for abelian étale sheaves obtained by passing to the direct limit over all étale coverings; see [**Mil80**, III.2]. For a quasi-compact scheme such that each finite subset in it is contained in an open affine subset (for example, a quasi-projective variety), Čech cohomology groups coincide with cohomology groups of abelian étale sheaves defined as above; see [**Mil80**, Theorem III.2.17]. We will say that such schemes are *good*.

Cohomology of an abelian étale sheaf on the spectrum of a field K is canonically isomorphic to Galois cohomology of the corresponding abelian G_K-module, as shown in [**Dan96**, 4.4.3]. The direct limit of Čech complexes over all étale coverings corresponds to the standard complex related to group cohomology (see Definition 1.1.3).

It follows from Artin's lemma that for smooth complex varieties, étale cohomology with finite constant coefficients coincides with Betti cohomology ([**Dan96**, 4.6.5], [**Mil80**, Theorem III.3.12]).

For any quasi-coherent sheaf $\mathcal{F}$ on X, there is a canonical isomorphism

$$H^i_{Zar}(X, \mathcal{F}) \cong H^i_{\acute{e}t}(X, \mathcal{F}_{\acute{e}t}),$$

where the subscript Zar stands for cohomology in Zariski topology. One can find details in [**Dan96**, 4.4.4], [**Mil80**, Proposition III.3.7], and [**Mil80**, Remark III.3.8].

Étale cohomology of a smooth quasi-projective variety with coefficients in the constant sheaf $\mathbb{Q}$ can be shown to be trivial in positive degrees: this follows from analogous statements on Zariski cohomology and on cohomology of profinite groups, from a computation of étale cohomology with the help of Čech cohomology, and from the Hochschild–Serre spectral sequence; see [**Mil80**, Theorem III.2.20]. Therefore one usually considers étale cohomology with coefficients in torsion sheaves. Let l be a prime number which does not divide the characteristic of a scheme X. Given a positive integer n, put

$$\mu_l^{\otimes -n} = \mathrm{Hom}(\mu_l^{\otimes n}, \mathbb{Z}/l\mathbb{Z})$$

and define *l-adic cohomology* by the formula

$$H^i_{\acute{e}t}\big(X, \mathbb{Z}_l(n)\big) = \varprojlim_r H^i_{\acute{e}t}\big(X, \mu_{l^r}^{\otimes n}\big), \quad H^i_{\acute{e}t}\big(X, \mathbb{Q}_l(n)\big) = \mathbb{Q}_l \otimes_{\mathbb{Z}_l} H^i_{\acute{e}t}\big(X, \mathbb{Z}_l(n)\big)\,;$$

see [**Dan96**, 4.7.1]. One has the pull-back map for l-adic cohomology. For a smooth projective variety X of dimension d over a separably closed field, one has Poincaré duality, that is, a non-degenerate pairing

$$H^i_{\acute{e}t}\big(X, \mathbb{Q}_l(n)\big) \otimes_{\mathbb{Q}_l} H^{2d-i}_{\acute{e}t}\big(X, \mathbb{Q}_l(d-n)\big) \to \mathbb{Q}_l\,;$$

see [**Dan96**, 4.7.5] or [**Mil80**, Theorem VI.11.1]. One has a push-forward map (Gysin homomorphism) for l-adic cohomology that satisfies the projection formula; see [**Dan96**, 4.7.6]. This implies an analog of the Lefschetz trace formula from topology; see [**Dan96**, 4.7.8]. One can also consider cohomology with compact support and with coefficients in a constructible l-adic sheaf. Such cohomology has many properties similar to those of cohomology of sheaves on manifolds with classical topology ([**Dan96**, 4.7], [**Mil80**, Chapter VI]). Altogether this allowed P. Deligne to prove Weil conjectures on zeta functions of varieties over finite fields using l-adic cohomology. See the details in [**Del74**], [**Del80**], [**Dan96**, 4.7.9, 4.7.10, 4.8], and [**Mil80**, VI.12, VI.13].

A.4. First étale cohomology with non-abelian coefficients

We address the following question: let $U \to X$ be a (finite) étale covering and let B be some algebro-geometric object over U; what conditions imply that B is a pull-back of an object A over X? How can we describe all such A? It turns out that the main definitions and constructions from Chapter 2 can be generalized from the case of the spectrum of a field to the case of an arbitrary scheme X. One can read about this in detail in [**Gro95c**] and also in the survey [**Vis05**].

So, let X be a scheme and $\mathcal{M}$ a category fibered over étale schemes over X; that is, for each étale morphism $U \to X$ one has a category $\mathcal{M}(U)$, and for each morphism $q\colon V \to U$ between étale schemes over X one has a pull-back functor

$$q^*\colon \mathcal{M}(U) \to \mathcal{M}(V)\,.$$

We require the pull-back functors to be compatible with compositions of morphisms between étale schemes over X. In light of the above questions, only isomorphisms between objects are important. Thus we also assume that for any U, the category $\mathcal{M}(U)$ is a *groupoid*, that is, all morphisms in $\mathcal{M}(U)$ are isomorphisms. For example, $\mathcal{M}(U)$ can be the category of vector bundles over U or the category of schemes of finite type over U (in both cases, for morphisms we take only isomorphisms).

For simplicity, we will consider only finite étale coverings (this is enough when X is quasi-compact). Let $f\colon U \to X$ be an étale covering and let

$$p_i\colon U \times_X U \to U, \quad p_{ij}\colon U \times_X U \times_X U \to U \times_X U$$

denote the corresponding projections. We say that one has *descent data* on an object B in $\mathcal{M}(U)$ if one has fixed an isomorphism $\rho\colon p_1^*B \xrightarrow{\sim} p_2^*B$ such that

$$p_{13}^*(\rho) = p_{23}^*(\rho) \circ p_{12}^*(\rho)\colon p_1^*B \to p_3^*B\,.$$

For example, for any object A in $\mathcal{M}(X)$, there are canonical descent data on f^*A. We say that $\mathcal{M}$ *satisfies descent* if the pull-back functor f^* gives an equivalence between the category $\mathcal{M}(X)$ and the category of objects in $\mathcal{M}(U)$ with descent data. If this condition is satisfied for any étale scheme V over X, not just for X, then $\mathcal{M}$ is called a *stack*. In this case, one can say that $\mathcal{M}$ satisfies the sheaf condition in the étale topology.

Given an object A in $\mathcal{M}(X)$, we have an étale presheaf of groups $\underline{\mathrm{Aut}}(A)$ defined by the formula $V \mapsto \mathrm{Aut}(q^*A)$ for an étale morphism $q\colon V \to X$. As in Exercise 2.2.1(i), there is a bijection between the set of isomorphism classes of descent data on f^*A and the set $\check{H}^1_{ét}\big(U/X, \underline{\mathrm{Aut}}(A)\big)$, where the first Čech cohomology with non-abelian coefficients generalizes the definition from Exercise 1.1.7(i); see [**Mil80**, III.4]. Furthermore, as in Exercise 2.2.1(iii), if $\mathcal{M}$ satisfies descent, then the set $\check{H}^1_{ét}\big(U/X, \underline{\mathrm{Aut}}(A)\big)$ is bijective with the set of isomorphism classes of U-forms of A, that is, objects A' in $\mathcal{M}(X)$ such that $f^*A \cong f^*A'$ *([**Mil80**, III.4], [**Dan96**, 4.4.5]).

Note that if $\mathcal{M}$ is a stack (which is the case in most reasonable examples), then $\underline{\mathrm{Aut}}(A)$ is a sheaf in étale topology. Similarly to Exercise 1.1.7(ii) and (iii), there is an exact sequence of pointed sets associated with the short exact sequence of sheaves of non-abelian groups on a good scheme; see Step 3 in the proof of Theorem IV.2.5 in [**Mil80**].

The category of quasi-coherent sheaves satisfies descent and is a stack; see, for example, [**Mil80**, Remark I.2.21] or [**Mil80**, Proposition I.2.22]. The proof of this fact follows the method of solution of Exercise 2.1.3. In particular, the category of vector bundles satisfies descent, whence it follows that the canonical map

$$\check{H}^1_{Zar}(X, \mathrm{GL}_n) \to \check{H}^1_{ét}(X, \mathrm{GL}_n)$$

is a bijection; see [**Mil80**, Proposition III.4.9]. For example, there is an isomorphism

$$\mathrm{Pic}(X) \cong H^1_{ét}(X, \mathbb{G}_m)\,.$$

The following categories also satisfy descent (and are stacks): the category of affine schemes ([**Mil80**, Theorem I.2.23]), the category of torsors under a smooth affine group scheme ([**Mil80**, Corollary III.4.7], [**Mil80**, Remark III.4.8]), and the category of schemes with a relatively ample sheaf ([**SGA71**, Proposition VIII.7.8]). In particular, the last example implies that the category of Severi–Brauer schemes, that is, families of Severi–Brauer varieties, satisfies descent and there is a bijection between the set of isomorphism classes of Severi–Brauer schemes and the set $\check{H}^1_{ét}(X, \mathrm{PGL}_n)$; see [**Mil80**, III.4]. Nevertheless, not all natural categories of schemes satisfy descent: for example, the category of elliptic schemes is not a stack as shown in [**Vis05**, Example 4.39]; this is the case because there exist non-quasi-projective varieties that do not admit quotients over free actions of finite groups in the category of schemes. This phenomenon leads to the notion of an algebraic space.

A.5. Kummer sequence

Suppose that a prime number l does not divide the characteristic of X and let r be a positive integer. There is an exact sequence of étale sheaves, called the *Kummer sequence* ([**Mil80**, III.4], [**Dan96**, 4.4.6]):

$$1 \to \mu_{l^r} \to \mathbb{G}_m \xrightarrow{l^r} \mathbb{G}_m \to 1\,. \tag{A.1}$$

This sequence is an analog of the exponential exact sequence of sheaves on complex manifolds. It generalizes the Kummer exact sequence in Exercise 2.4.2(i). As was mentioned in §A.4, there is an isomorphism

$$\mathrm{Pic}(X) \cong H^1_{\acute{e}t}(X, \mathbb{G}_m)\,.$$

The corresponding long exact sequence leads to short exact sequences

$$1 \to \Gamma(X, \mathcal{O}_X^*)/l^r \to H^1_{\acute{e}t}(X, \mu_{l^r}) \to \mathrm{Pic}(X)_{l^r} \to 0\,, \tag{A.2}$$

$$0 \to \mathrm{Pic}(X)/l^r \to H^2_{\acute{e}t}(X, \mu_{l^r}) \to H^2_{\acute{e}t}(X, \mathbb{G}_m)_{l^r} \to 0\,. \tag{A.3}$$

In particular, if X is a complete scheme over a separably closed field, then (A.2) implies the isomorphism

$$H^1_{\acute{e}t}(X, \mu_{l^r}) \cong \mathrm{Pic}(X)_{l^r}\,.$$

Also note that the exact sequence (A.3) is a generalization of the isomorphism from Exercise 3.1.10(iii).

For an arbitrary scheme X, upon passing to the direct limit in (A.3), we obtain an exact sequence

$$0 \to \mathrm{Pic}(X) \otimes_{\mathbb{Z}} \mathbb{Q}_l/\mathbb{Z}_l \to H^2_{\acute{e}t}(X, \mu_{l^\infty}) \to H^2_{\acute{e}t}(X, \mathbb{G}_m)_{l^\infty} \to 0\,, \tag{A.4}$$

where $A_{l^\infty} = \bigcup_{r\geqslant 1} A_{l^r}$ for an abelian group A and we define $\mu_{l^\infty} = \bigcup_{r\geqslant 1} \mu_{l^r}$. If the characteristic of X is zero, then there is an isomorphism

$$H^2_{\acute{e}t}(X, \mathbb{G}_m)_{tors} \cong \mathrm{Coker}\Big(\mathrm{Pic}(X) \otimes_{\mathbb{Z}} \mathbb{Q} \to H^2_{\acute{e}t}(X, \mu_\infty)\Big)\,, \tag{A.5}$$

where $\mu_\infty = \prod_l \mu_{l^\infty}$. Thus one can say that $H^2_{\acute{e}t}(X, \mathbb{G}_m)_{tors}$ is the group of transcendental classes in second étale cohomology.

An important property of the group $H^2_{\acute{e}t}(X, \mathbb{G}_m)_{tors}$ is that it is a stable birational invariant for smooth projective varieties X over a field k of characteristic zero. Indeed, one easily deduces from (A.5) that the group $H^2_{\acute{e}t}(X, \mathbb{G}_m)_{tors}$ does not change under blow-ups with smooth centers or under the product with $\mathbb{P}^1$. Now let $f\colon X \dashrightarrow Y$ be a birational map between smooth projective varieties over k. By Hironaka's desingularization theorem, there is a smooth projective variety $\widetilde{X}$ and birational morphisms π and $\tilde{f}$ which form a commutative diagram

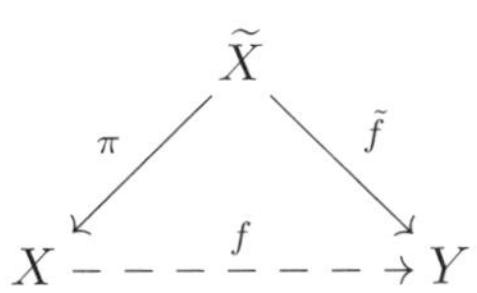

Moreover, the morphism π is a sequence of blow-ups with smooth centers. In particular, there is an isomorphism

$$\pi^*\colon H^2_{\acute{e}t}(X, \mathbb{G}_m)_{tors} \xrightarrow{\sim} H^2_{\acute{e}t}(\widetilde{X}, \mathbb{G}_m)_{tors}\,,$$

and there is a map

$$(\pi^*)^{-1} \circ \tilde{f}^*\colon H^2_{\acute{e}t}(Y, \mathbb{G}_m)_{tors} \to H^2_{\acute{e}t}(X, \mathbb{G}_m)_{tors}\,.$$

Similarly, regularizing the birational map f^{-1}, we obtain a map

$$H^2_{\acute{e}t}(X, \mathbb{G}_m)_{tors} \to H^2_{\acute{e}t}(Y, \mathbb{G}_m)_{tors}\,.$$

It is easy to check that these maps are mutually inverse.

A.6. Brauer group

The main definitions and constructions from Chapter 3 can be generalized from fields to schemes. Using the above theory of descent and proceeding as in parts (i), (ii), and (iii) of Exercise 3.1.2, one can show that there is a canonical embedding of groups (see [**Mil80**, Theorem IV.2.5])

$$\lambda\colon \operatorname{Br}(X) \hookrightarrow H^2_{ét}(X, \mathbb{G}_m),$$

where $\operatorname{Br}(X)$ is defined in terms of Azumaya algebras; see Definition 4.4.1 (for this approach one should also require that the scheme X be good).

In the same way as in Exercise 3.1.10(ii), one can show that the group $\operatorname{Br}(X)$ is torsion; see [**Mil80**, Proposition IV.2.7] (for the case of algebras of rank not coprime with the characteristic, one uses flat topology). It is important to mention that Exercise 3.1.2(iv) does not generalize to the case of arbitrary schemes, that is, the embedding λ is not necessarily an isomorphism. In particular, [**Gro95b**, Remarque 1.11] gives an example of a singular surface X such that the group $H^2_{ét}(X, \mathbb{G}_m)$ has an element of infinite order. However, recently de Jong [**dJ03**] proved that for any scheme X that admits an ample sheaf, the above embedding λ gives an isomorphism

$$\lambda\colon \operatorname{Br}(X) \xrightarrow{\sim} H^2_{ét}(X, \mathbb{G}_m)_{tors}.$$

This fact was also proved earlier by Gabber using a different method.

For a regular irreducible variety X over a field k, it is not hard to show that the natural map

$$\nu\colon H^2_{ét}(X, \mathbb{G}_m) \to \operatorname{Br}\big(k(X)\big)$$

is injective; see [**Mil80**, Example III.2.22]. In particular, since the group $\operatorname{Br}\big(k(X)\big)$ is torsion (see Exercise 3.1.10(o)), for such X there is an equality

$$H^2_{ét}(X, \mathbb{G}_m) = H^2_{ét}(X, \mathbb{G}_m)_{tors}. \tag{A.6}$$

Thus the arguments in §A.5 imply that the group $H^2_{ét}(X, \mathbb{G}_m)$ is a stable birational invariant for smooth projective varieties over a field of characteristic zero.

For a smooth variety X over a field of characteristic zero, the map ν provides an isomorphism

$$\nu\colon H^2_{ét}(X, \mathbb{G}_m) \xrightarrow{\sim} \operatorname{Br}^{nr}(X).$$

This follows from purity for étale cohomology with finite coefficients and from the Kummer exact sequence in §A.5. One can find details in [**CT95**, §3.4]. This implies that the group $\operatorname{Br}^{nr}(X)$ is a stable birational invariant for smooth projective varieties over a field of characteristic zero.

A famous conjecture of M. Artin claims that the Brauer group $\operatorname{Br}(X)$ is finite for any proper scheme X over $\mathbb{Z}$; see [**Mil80**, Question IV.2.19].

A.7. The case of a complex algebraic variety

Let X be a smooth complex projective variety. We consider the topology on the set of its $\mathbb{C}$-points $X(\mathbb{C})$ induced by the classical topology on $\mathbb{C}$. By $H^i\big(X(\mathbb{C}), -\big)$ we denote Betti cohomology groups of the topological space $X(\mathbb{C})$, or cohomology groups with coefficients in a sheaf on this topological space.

The choice of a root of unity defines an isomorphism

$$H^2_{ét}(X, \mu_\infty) \cong H^2\big(X(\mathbb{C}), \mathbb{Q}/\mathbb{Z}\big). \tag{A.7}$$

It follows from (A.5), (A.6), (A.7), and de Jong's result mentioned in §A.6 that there is an isomorphism

$$\mathrm{Br}(X) \cong \mathrm{Coker}\left(\mathrm{Pic}(X) \otimes_{\mathbb{Z}} \mathbb{Q} \xrightarrow{r} H^2(X(\mathbb{C}), \mathbb{Q}/\mathbb{Z})\right).$$

On the other hand, the exact sequence of constant sheaves in classical topology,

$$0 \to \mathbb{Z} \to \mathbb{Q} \to \mathbb{Q}/\mathbb{Z} \to 0,$$

yields the exact sequence

$$H^2(X(\mathbb{C}), \mathbb{Z}) \to H^2(X(\mathbb{C}), \mathbb{Q}) \to H^2(X(\mathbb{C}), \mathbb{Q}/\mathbb{Z}) \to H^3(X(\mathbb{C}), \mathbb{Z})_{tors} \to 0.$$

Since the map r factors through $H^2(X(\mathbb{C}), \mathbb{Q})$, there is a short exact sequence

$$0 \to \mathrm{Coker}\Big((\mathrm{Pic}(X) \otimes_{\mathbb{Z}} \mathbb{Q}) \oplus H^2(X(\mathbb{C}), \mathbb{Z}) \to H^2(X(\mathbb{C}), \mathbb{Q})\Big) \\ \to \mathrm{Br}(X) \to H^3(X(\mathbb{C}), \mathbb{Z})_{tors} \to 0.$$

Finally, the (proven) Hodge conjecture for divisors asserts the equality

$$\left(H^0(X, \Omega^2_X) \oplus H^1(X, \Omega^1_X)\right) \cap H^2(X(\mathbb{C}), \mathbb{Q}) = \mathrm{Im}\left(\mathrm{Pic}(X) \otimes_{\mathbb{Z}} \mathbb{Q} \to H^2(X(\mathbb{C}), \mathbb{C})\right),$$

where Ω^i_X denotes the sheaf of algebraic i-forms on the variety X. Hence the natural projection $H^2(X(\mathbb{C}), \mathbb{C}) \to H^2(X, \mathcal{O}_X)$ induces an isomorphism

$$\mathrm{Coker}\Big((\mathrm{Pic}(X) \otimes_{\mathbb{Z}} \mathbb{Q}) \oplus H^2(X(\mathbb{C}), \mathbb{Z}) \to H^2(X(\mathbb{C}), \mathbb{Q})\Big) \\ \cong \mathrm{Coker}\left(H^2(X(\mathbb{C}), \mathbb{Z}) \to H^2(X, \mathcal{O}_X)\right)_{tors}.$$

Thus, there is an exact sequence

(A.8)

$$0 \to \mathrm{Coker}\left(H^2(X(\mathbb{C}), \mathbb{Z}) \to H^2(X, \mathcal{O}_X)\right)_{tors} \to \mathrm{Br}(X) \to H^3(X(\mathbb{C}), \mathbb{Z})_{tors} \to 0.$$

In particular, if the group $H^3(X(\mathbb{C}), \mathbb{Z})_{tors}$ is non-trivial, then the group $\mathrm{Br}(X)$ is non-trivial as well. Using (A.8) and the exponential exact sequence, one can show that the natural map

$$\mathrm{Br}(X) \to H^2(X(\mathbb{C}), \mathcal{A}^*_X)_{tors}$$

is an isomorphism, where $\mathcal{A}^*_X$ is the sheaf of invertible holomorphic functions on the complex manifold $X(\mathbb{C})$.

Bibliography

[ABGV11] Asher Auel, Eric Brussel, Skip Garibaldi, and Uzi Vishne, *Open problems on central simple algebras*, Transform. Groups **16** (2011), no. 1, 219–264. MR2785502

[ABP17] A. Auel, Ch. Böhning, and A. Pirutka. *Stable rationality of quadric and cubic surface bundle fourfolds*, arXiv:1710.07270, 2017.

[ABvBP16] A. Auel, Ch. Böhning, H.-Ch. Graf von Bothmer, and A. Pirutka. *Conic bundles with nontrivial unramified Brauer group over threefolds*, preprint arXiv:1610.04995, 2016.

[AM72] M. Artin and D. Mumford, *Some elementary examples of unirational varieties which are not rational*, Proc. London Math. Soc. (3) **25** (1972), 75–95. MR0321934

[AM96] Paul S. Aspinwall and David R. Morrison, *Stable singularities in string theory*, Comm. Math. Phys. **178** (1996), no. 1, 115–134. With an appendix by Mark Gross. MR1387944

[Ami55] S. A. Amitsur, *Generic splitting fields of central simple algebras*, Ann. of Math. (2) **62** (1955), 8–43. MR0070624

[AO16] H. Ahmadinezhad and T. Okada. *Stable rationality of higher dimensional conic bundles*, preprint arXiv:1612.04206, 2016.

[Ara75] Jón Kr. Arason, *Cohomologische invarianten quadratischer Formen* (French), J. Algebra **36** (1975), no. 3, 448–491. MR0389761

[Art82] M. Artin, *Left ideals in maximal orders*, Brauer groups in ring theory and algebraic geometry (Wilrijk, 1981), Lecture Notes in Math., vol. 917, Springer, Berlin-New York, 1982, pp. 182–193. MR657429

[Aso13] Aravind Asok, *Rationality problems and conjectures of Milnor and Bloch-Kato*, Compos. Math. **149** (2013), no. 8, 1312–1326. MR3103066

[BB13] Fedor Bogomolov and Christian Böhning, *Isoclinism and stable cohomology of wreath products*, Birational geometry, rational curves, and arithmetic, Simons Symp., Springer, Cham, 2013, pp. 57–76. MR3114923

[BCTSSD85] Arnaud Beauville, Jean-Louis Colliot-Thélène, Jean-Jacques Sansuc, and Peter Swinnerton-Dyer, *Variétés stablement rationnelles non rationnelles* (French), Ann. of Math. (2) **121** (1985), no. 2, 283–318. MR786350

[Bea96] Arnaud Beauville, *Complex algebraic surfaces*, 2nd ed., London Mathematical Society Student Texts, vol. 34, Cambridge University Press, Cambridge, 1996. Translated from the 1978 French original by R. Barlow, with assistance from N. I. Shepherd-Barron and M. Reid. MR1406314

[Bea16] Arnaud Beauville, *A very general sextic double solid is not stably rational*, Bull. Lond. Math. Soc. **48** (2016), no. 2, 321–324. MR3483069

[Ber10] Grégory Berhuy, *An introduction to Galois cohomology and its applications*, London Mathematical Society Lecture Note Series, vol. 377, Cambridge University Press, Cambridge, 2010. With a foreword by Jean-Pierre Tignol. MR2723693

[Bir62] B. J. Birch, *Forms in many variables*, Proc. Roy. Soc. Ser. A **265** (1961/1962), 245–263. MR0150129

[BLR90] Siegfried Bosch, Werner Lütkebohmert, and Michel Raynaud, *Néron models*, Ergebnisse der Mathematik und ihrer Grenzgebiete (3) [Results in Mathematics and Related Areas (3)], vol. 21, Springer-Verlag, Berlin, 1990. MR1045822

[BM13] T. Browning and L. Matthiesen. *Norm forms for arbitrary number fields as products of linear polynomials*, preprint arXiv:1307.7641, 2013.

[BMS14] T. D. Browning, L. Matthiesen, and A. N. Skorobogatov, *Rational points on pencils of conics and quadrics with many degenerate fibers*, Ann. of Math. (2) **180** (2014), no. 1, 381–402. MR3194818

[Bog87] F. A. Bogomolov, *The Brauer group of quotient spaces of linear representations* (Russian), Izv. Akad. Nauk SSSR Ser. Mat. **51** (1987), no. 3, 485–516, 688; English transl., Math. USSR-Izv. **30** (1988), no. 3, 455–485. MR903621

[Bor96] Mikhail Borovoi, *The Brauer-Manin obstructions for homogeneous spaces with connected or abelian stabilizer*, J. Reine Angew. Math. **473** (1996), 181–194. MR1390687

[Bou58] N. Bourbaki, *Éléments de mathématique. 23. Première partie: Les structures fondamentales de l'analyse. Livre II: Algèbre. Chapitre 8: Modules et anneaux semi-simples* (French), Actualités Sci. Ind. no. 1261, Hermann, Paris, 1958. MR0098114

[BP11] Fedor Bogomolov and Tihomir Petrov, *Unramified cohomology of alternating groups*, Cent. Eur. J. Math. **9** (2011), no. 5, 936–948. MR2824438

[BP13] Fedor Bogomolov and Yuri Prokhorov, *On stable conjugacy of finite subgroups of the plane Cremona group, I*, Cent. Eur. J. Math. **11** (2013), no. 12, 2099–2105. MR3111709

[BT17] F. Bogomolov and Yu. Tschinkel. *Noether's problem and descent*, arXiv:1711.09465, 2017.

[BvB16] Ch. Böhning and H.-Ch. Graf von Bothmer. *On stable rationality of some conic bundles and moduli spaces of Prym curves*, arXiv:1605.03029, 2016.

[Cao17] Y. Cao. *Sous-groupe de Brauer invariant et obstruction de descente itérée*, arXiv:1704.05425, 2017.

[CF67] J. W. S. Cassels and A. Fröhlich (Eds.). *Algebraic number theory*, Academic Press, London, 1967.

[CG66] J. W. S. Cassels and M. J. T. Guy, *On the Hasse principle for cubic surfaces*, Mathematika **13** (1966), 111–120. MR0211966

[CHHK15] Huah Chu, Akinari Hoshi, Shou-Jen Hu, and Ming-chang Kang, *Noether's problem for groups of order 243*, J. Algebra **442** (2015), 233–259. MR3395061

[CHKK10] Huah Chu, Shou-Jen Hu, Ming-chang Kang, and Boris E. Kunyavskii, *Noether's problem and the unramified Brauer group for groups of order 64*, Int. Math. Res. Not. IMRN **12** (2010), 2329–2366. MR2652224

[CHKP08] Huah Chu, Shou-Jen Hu, Ming-chang Kang, and Y. G. Prokhorov, *Noether's problem for groups of order 32*, J. Algebra **320** (2008), no. 7, 3022–3035. MR2442008

[CK01] Huah Chu and Ming-chang Kang, *Rationality of p-group actions*, J. Algebra **237** (2001), no. 2, 673–690. MR1816710

[Cla06] Pete L. Clark, *On elementary equivalence, isomorphism and isogeny* (English, with English and French summaries), J. Théor. Nombres Bordeaux **18** (2006), no. 1, 29–58. MR2245874

[Cor05] Patrick Corn, *Del Pezzo surfaces of degree 6*, Math. Res. Lett. **12** (2005), no. 1, 75–84. MR2122732

[CT95] J.-L. Colliot-Thélène, *Birational invariants, purity and the Gersten conjecture*, K-theory and algebraic geometry: connections with quadratic forms and division algebras (Santa Barbara, CA, 1992), Proc. Sympos. Pure Math., vol. 58, Amer. Math. Soc., Providence, RI, 1995, pp. 1–64. MR1327280

[CT03] Jean-Louis Colliot-Thélène, *Points rationnels sur les fibrations* (French), Higher dimensional varieties and rational points (Budapest, 2001), Bolyai Soc. Math. Stud., vol. 12, Springer, Berlin, 2003, pp. 171–221. MR2011747

[CT17] J.-L. Colliot-Thélène. *Surfaces stablement rationnelles sur un corps quasi-fini*, arXiv:1711.09595, 2017.

[CTO89] Jean-Louis Colliot-Thélène and Manuel Ojanguren, *Variétés unirationnelles non rationnelles: au-delà de l'exemple d'Artin et Mumford* (French), Invent. Math. **97** (1989), no. 1, 141–158. MR999316

[CTP16a] Zh.-L. Kol′ë-Telën and E. V. Piryutko, *Cyclic covers that are not stably rational* (Russian, with Russian summary), Izv. Ross. Akad. Nauk Ser. Mat. **80** (2016), no. 4, 35–48; English transl., Izv. Math. **80** (2016), no. 4, 665–677. MR3535357

[CTP16b] Jean-Louis Colliot-Thélène and Alena Pirutka, *Hypersurfaces quartiques de dimension 3: non-rationalité stable* (French, with English and French summaries), Ann. Sci. Éc. Norm. Supér. (4) **49** (2016), no. 2, 371–397. MR3481353

[CTPS16] Jean-Louis Colliot-Thélène, Ambrus Pál, and Alexei N. Skorobogatov, *Pathologies of the Brauer-Manin obstruction*, Math. Z. **282** (2016), no. 3-4, 799–817. MR3473644

[CTS87] Jean-Louis Colliot-Thélène and Jean-Jacques Sansuc, *La descente sur les variétés rationnelles. II* (French), Duke Math. J. **54** (1987), no. 2, 375–492. MR899402

[CTS07] Jean-Louis Colliot-Thélène and Jean-Jacques Sansuc, *The rationality problem for fields of invariants under linear algebraic groups (with special regards to the Brauer group)*, Algebraic groups and homogeneous spaces, Tata Inst. Fund. Res. Stud. Math., vol. 19, Tata Inst. Fund. Res., Mumbai, 2007, pp. 113–186. MR2348904

[CTSD94] Jean-Louis Colliot-Thélène and Peter Swinnerton-Dyer, *Hasse principle and weak approximation for pencils of Severi-Brauer and similar varieties*, J. Reine Angew. Math. **453** (1994), 49–112. MR1285781

[CTSSD87a] Jean-Louis Colliot-Thélène, Jean-Jacques Sansuc, and Peter Swinnerton-Dyer, *Intersections of two quadrics and Châtelet surfaces. I*, J. Reine Angew. Math. **373** (1987), 37–107. MR870307

[CTSSD87b] Jean-Louis Colliot-Thélène, Jean-Jacques Sansuc, and Peter Swinnerton-Dyer, *Intersections of two quadrics and Châtelet surfaces. I*, J. Reine Angew. Math. **373** (1987), 37–107. MR870307

[CTSSD87c] Jean-Louis Colliot-Thélène, Jean-Jacques Sansuc, and Peter Swinnerton-Dyer, *Intersections of two quadrics and Châtelet surfaces. II*, J. Reine Angew. Math. **374** (1987), 72–168. MR876222

[Dan96] V. I. Danilov, *Cohomology of algebraic varieties [MR1060324 (91f:14016)]*, Algebraic geometry, II, Encyclopaedia Math. Sci., vol. 35, Springer, Berlin, 1996, pp. 1–125, 255–262. MR1392958

[Dav63] H. Davenport, *Cubic forms in sixteen variables*, Proc. Roy. Soc. Ser. A **272** (1963), 285–303. MR0155800

[Del74] Pierre Deligne, *La conjecture de Weil. I* (French), Inst. Hautes Études Sci. Publ. Math. **43** (1974), 273–307. MR0340258

[Del80] Pierre Deligne, *La conjecture de Weil. II* (French), Inst. Hautes Études Sci. Publ. Math. **52** (1980), 137–252. MR601520

[Dem09] Cyril Demarche, *Obstruction de descente et obstruction de Brauer-Manin étale* (French, with English and French summaries), Algebra Number Theory **3** (2009), no. 2, 237–254. MR2491944

[DF84] Frank R. DeMeyer and Timothy J. Ford, *On the Brauer group of surfaces*, J. Algebra **86** (1984), no. 1, 259–271. MR727378

[DG94] Igor Dolgachev and Mark Gross, *Elliptic threefolds. I. Ogg-Shafarevich theory*, J. Algebraic Geom. **3** (1994), no. 1, 39–80. MR1242006

[dJ03] A. J. de Jong. *A result of Gabber*, `http://www.math.columbia.edu/~dejong/papers/2-gabber.pdf`, 2003.

[dJ04] A. J. de Jong, *The period-index problem for the Brauer group of an algebraic surface*, Duke Math. J. **123** (2004), no. 1, 71–94. MR2060023

[dlBBP12] Régis de la Bretèche, Tim Browning, and Emmanuel Peyre, *On Manin's conjecture for a family of Châtelet surfaces*, Ann. of Math. (2) **175** (2012), no. 1, 297–343. MR2874644

[Dol12] Igor V. Dolgachev, *Classical algebraic geometry: a modern view*, Cambridge University Press, Cambridge, 2012. MR2964027

[EKM08] Richard Elman, Nikita Karpenko, and Alexander Merkurjev, *The algebraic and geometric theory of quadratic forms*, American Mathematical Society Colloquium Publications, vol. 56, American Mathematical Society, Providence, RI, 2008. MR2427530

[EM75] Shizuo Endô and Takehiko Miyata, *On a classification of the function fields of algebraic tori*, Nagoya Math. J. **56** (1975), 85–104. MR0364203

[Esn03] Hélène Esnault, *Varieties over a finite field with trivial Chow group of 0-cycles have a rational point*, Invent. Math. **151** (2003), no. 1, 187–191. MR1943746

[FK88] Eberhard Freitag and Reinhardt Kiehl, *Étale cohomology and the Weil conjecture*, Ergebnisse der Mathematik und ihrer Grenzgebiete (3) [Results in Mathematics and Related Areas (3)], vol. 13, Springer-Verlag, Berlin, 1988. Translated from the German by Betty S. Waterhouse and William C. Waterhouse; With an historical introduction by J. A. Dieudonné. MR926276

[Gab98] Ofer Gabber, *A note on the unramified Brauer group and purity*, Manuscripta Math. **95** (1998), no. 1, 107–115. MR1492372

[Gro68] Alexander Grothendieck, *Le groupe de Brauer. III. Exemples et compléments* (French), Dix exposés sur la cohomologie des schémas, Adv. Stud. Pure Math., vol. 3, North-Holland, Amsterdam, 1968, pp. 88–188. MR244271

[Gro95a] Alexander Grothendieck, *Le groupe de Brauer. I. Algèbres d'Azumaya et interprétations diverses [MR0244269 (39 #5586a)]* (French), Séminaire Bourbaki, Vol. 9 Exp. No. 290, Soc. Math. France, Paris, 1995, pp. 199–219. MR1608798

[Gro95b] Alexander Grothendieck, *Le groupe de Brauer. II. Théorie cohomologique [MR0244270 (39 #5586b)]* (French), Séminaire Bourbaki, Vol. 9, Exp. No. 297, Soc. Math. France, Paris, 1995, pp. 287–307. MR1608805

[Gro95c] Alexander Grothendieck, *Technique de descente et théorèmes d'existence en géométrie algébrique. I. Généralités. Descente par morphismes fidèlement plats* (French), Séminaire Bourbaki, Vol. 5, Exp. No. 190, Soc. Math. France, Paris, 1995, pp. 299–327. MR1603475

[GS06] Philippe Gille and Tamás Szamuely, *Central simple algebras and Galois cohomology*, Cambridge Studies in Advanced Mathematics, vol. 101, Cambridge University Press, Cambridge, 2006. MR2266528

[Har77] Robin Hartshorne, *Algebraic geometry*, Graduate Texts in Mathematics, vol. 52, Springer-Verlag, New York-Heidelberg, 1977. MR0463157

[Har94] David Harari, *Méthode des fibrations et obstruction de Manin* (French), Duke Math. J. **75** (1994), no. 1, 221–260. MR1284820

[Har95] Joe Harris, *Algebraic geometry: a first course*, Graduate Texts in Mathematics, vol. 133, Springer-Verlag, New York, 1995. Corrected reprint of the 1992 original. MR1416564

[Har02] David Harari, *Groupes algébriques et points rationnels* (French, with English and French summaries), Math. Ann. **322** (2002), no. 4, 811–826. MR1905103

[HB83] D. R. Heath-Brown, *Cubic forms in ten variables*, Proc. London Math. Soc. (3) **47** (1983), no. 2, 225–257. MR703978

[HK10] Akinari Hoshi and Ming-Chang Kang, *Twisted symmetric group actions*, Pacific J. Math. **248** (2010), no. 2, 285–304. MR2741249

[HK11] A. Hoshi and M. Kang. *Unramified Brauer groups for groups of order p^5*, arXiv:1109.2966, 2011.

[HKK13] Akinari Hoshi, Ming-Chang Kang, and Boris E. Kunyavskii, *Noether's problem and unramified Brauer groups*, Asian J. Math. **17** (2013), no. 4, 689–713. MR3152260

[HKT16] Brendan Hassett, Andrew Kresch, and Yuri Tschinkel, *Stable rationality and conic bundles*, Math. Ann. **365** (2016), no. 3-4, 1201–1217. MR3521088

[HKY16] Akinari Hoshi, Ming-chang Kang, and Aiichi Yamasaki, *Degree three unramified cohomology groups*, J. Algebra **458** (2016), 120–133. MR3500770

[Hoo13] Christopher Hooley, *On nonary cubic forms: IV*, J. Reine Angew. Math. **680** (2013), 23–39. MR3100951

[Hos16] Akinari Hoshi, *Birational classification of fields of invariants for groups of order 128*, J. Algebra **445** (2016), 394–432. MR3418064

[HPT16a] B. Hassett, A. Pirutka, and Yu. Tschinkel. *Stable rationality of quadric surface bundles over surfaces*, arXiv:1603.09262, 2016.

[HPT16b] B. Hassett, A. Pirutka, and Yu. Tschinkel. *A very general quartic double fourfold is not stably rational*, arXiv:1605.03220, 2016.

[HPT17] B. Hassett, A. Pirutka, and Yu. Tschinkel. *Intersections of three quadrics in $\mathbb{P}^7$*, arXiv:1706.01371, 2017.

[HS02] David Harari and Alexei N. Skorobogatov, *Non-abelian cohomology and rational points*, Compositio Math. **130** (2002), no. 3, 241–273. MR1887115

[HS14] Yonatan Harpaz and Alexei N. Skorobogatov, *Singular curves and the étale Brauer-Manin obstruction for surfaces* (English, with English and French summaries), Ann. Sci. Éc. Norm. Supér. (4) **47** (2014), no. 4, 765–778. MR3250063

[HSW14] Yonatan Harpaz, Alexei N. Skorobogatov, and Olivier Wittenberg, *The Hardy-Littlewood conjecture and rational points*, Compos. Math. **150** (2014), no. 12, 2095–2111. MR3292295

[HT04] Brendan Hassett and Yuri Tschinkel, *Universal torsors and Cox rings*, Arithmetic of higher-dimensional algebraic varieties (Palo Alto, CA, 2002), Progr. Math., vol. 226, Birkhäuser Boston, Boston, MA, 2004, pp. 149–173. MR2029868

[HT16] B. Hassett and Yu. Tschinkel. *On stable rationality of Fano threefolds and del Pezzo fibrations*, arXiv:1601.07074, 2016.

[Huh13] J. Huh. *A counterexample to the geometric Chevalley–Warning conjecture*, arXiv:1307.7765, 2013.

[HW16] Yonatan Harpaz and Olivier Wittenberg, *On the fibration method for zero-cycles and rational points*, Ann. of Math. (2) **183** (2016), no. 1, 229–295. MR3432584

[IKP14] Atanas Iliev, Ludmil Katzarkov, and Victor Przyjalkowski, *Double solids, categories and non-rationality*, Proc. Edinb. Math. Soc. (2) **57** (2014), no. 1, 145–173. MR3165018

[IM71] V. A. Iskovskih and Ju. I. Manin, *Three-dimensional quartics and counterexamples to the Lüroth problem* (Russian), Mat. Sb. (N.S.) **86(128)** (1971), 140–166. MR0291172

[Isk67] V. A. Iskovskih, *Rational surfaces with a pencil of rational curves* (Russian), Mat. Sb. (N.S.) **74(116)** (1967), 608–638. MR0220734

[Isk70] V. A. Iskovskih, *Rational surfaces with a sheaf of rational curves and with a positive square of canonical class* (Russian), Mat. Sb. (N.S.) **83(125)** (1970), 90–119. MR0271109

[Isk71a] V. A. Iskovskih, *Birational properties of a surface of degree* 4 *in* P_k^4 (Russian), Mat. Sb. (N.S.) **88(130)** (1971), 31–37. MR0299606

[Isk71b] V. A. Iskovskih, *A counterexample to the Hasse principle for systems of two quadratic forms in five variables* (Russian), Mat. Zametki **10** (1971), 253–257. MR0286743

[Isk96] V. A. Iskovskikh, *Factorization of birational mappings of rational surfaces from the point of view of Mori theory* (Russian), Uspekhi Mat. Nauk **51** (1996), no. 4(310), 3–72; English transl., Russian Math. Surveys **51** (1996), no. 4, 585–652. MR1422227

[IT91] V. A. Iskovskikh and S. L. Tregub, *Birational automorphisms of rational surfaces* (Russian), Izv. Akad. Nauk SSSR Ser. Mat. **55** (1991), no. 2, 254–281; English transl., Math. USSR-Izv. **38** (1992), no. 2, 251–275. MR1133298

[Izh04] Oleg T. Izhboldin, Bruno Kahn, Nikita A. Karpenko, and Alexander Vishik, *Geometric methods in the algebraic theory of quadratic forms*, Lecture Notes in Mathematics, vol. 1835, Springer-Verlag, Berlin, 2004. Edited by Jean-Pierre Tignol. MR2069219

[Jac75] Nathan Jacobson, *Lectures in abstract algebra. III. Theory of fields and Galois theory*, Graduate Texts in Mathematics, vol. 32, Springer-Verlag, New York-Heidelberg, 1975. Second corrected printing. MR0392906

[JM14] Urban Jezernik and Primož Moravec, *Bogomolov multipliers of groups of order 128*, Exp. Math. **23** (2014), no. 2, 174–180. MR3223772

[JM15] Urban Jezernik and Primož Moravec, *Universal commutator relations, Bogomolov multipliers, and commuting probability*, J. Algebra **428** (2015), 1–25. MR3314282

[Kah08] Bruno Kahn, *Formes quadratiques sur un corps* (French), Cours Spécialisés [Specialized Courses], vol. 15, Société Mathématique de France, Paris, 2008. MR2568030

[Kan06] Ming-Chang Kang, *Noether's problem for metacyclic p-groups*, Adv. Math. **203** (2006), no. 2, 554–567. MR2227733

[Kan09] Ming-chang Kang, *Retract rationality and Noether's problem*, Int. Math. Res. Not. IMRN **15** (2009), 2760–2788. MR2525840

[KK14] Ming-chang Kang and Boris Kunyavskiĭ, *The Bogomolov multiplier of rigid finite groups*, Arch. Math. (Basel) **102** (2014), no. 3, 209–218. MR3181710

[KO17] I. Krylov and T. Okada. *Stable rationality of del Pezzo fibrations of low degree over projective spaces*, arXiv:1701.08372, 2017.

[Kol96] János Kollár, *Rational curves on algebraic varieties*, Ergebnisse der Mathematik und ihrer Grenzgebiete. 3. Folge. A Series of Modern Surveys in Mathematics [Results in Mathematics and Related Areas. 3rd Series. A Series of Modern Surveys in Mathematics], vol. 32, Springer-Verlag, Berlin, 1996. MR1440180

[Kol16a] J. Kollár. *Severi–Brauer varieties; a geometric treatment*, arXiv:1606.04368, 2016.

[Kol16b] J. Kollár. *Symmetric powers of Severi–Brauer varieties*, arXiv:1603.02104, 2016.

[Kor15] Yitzhak Kornbluth, *Noether's problem for the five unsolved groups of order 64: a quadric model*, ProQuest LLC, Ann Arbor, MI, 2015. Thesis (Ph.D.)–New York University. MR3419255

[KP09] Ming-Chang Kang and Bernat Plans, *Reduction theorems for Noether's problem*, Proc. Amer. Math. Soc. **137** (2009), no. 6, 1867–1874. MR2480265

[Kra01] Daniel Reuben Krashen, *Birational isomorphisms between Severi-Brauer varieties*, ProQuest LLC, Ann Arbor, MI, 2001. Thesis (Ph.D.)–The University of Texas at Austin. MR2702739

[KS04] Daniel Krashen and David J. Saltman, *Severi-Brauer varieties and symmetric powers*, Algebraic transformation groups and algebraic varieties, Encyclopaedia Math. Sci., vol. 132, Springer, Berlin, 2004, pp. 59–70. MR2090670

[KT11] Andrew Kresch and Yuri Tschinkel, *Effectivity of Brauer-Manin obstructions on surfaces*, Adv. Math. **226** (2011), no. 5, 4131–4144. MR2770443

[KT17a] M. Kontsevich and Yu. Tschinkel. *Specialization of birational types*, arXiv:1708.05699, 2017.

[KT17b] A. Kresch and Yu. Tschinkel. *Stable rationality of Brauer–Severi surface bundles*, arXiv:1709.10151, 2017.

[Kun07] B. Kunyavskiĭ. *Algebraic tori — thirty years after* Algebra and number theory (Samara State Univ., May 2007), arXiv:0712.4061 [math.AG]

[Kun10] Boris Kunyavskiĭ, *The Bogomolov multiplier of finite simple groups*, Cohomological and geometric approaches to rationality problems, Progr. Math., vol. 282, Birkhäuser Boston, Inc., Boston, MA, 2010, pp. 209–217. MR2605170

[KW14] Ming-chang Kang and Baoshan Wang, *Rational invariants for subgroups of S_5 and S_7*, J. Algebra **413** (2014), 345–363. MR3216611

[KWZ15] Ming-chang Kang, Baoshan Wang, and Jian Zhou, *Invariants of wreath products and subgroups of S_6*, Kyoto J. Math. **55** (2015), no. 2, 257–279. MR3356073

[KZ12] Ming-Chang Kang and Jian Zhou, *Noether's problem for $\widehat{S}_4$ and $\widehat{S}_5$*, Pacific J. Math. **258** (2012), no. 2, 349–368. MR2981958

[KZ17] Huah Chu and Shang Huang, *Noether's problem on semidirect product groups*, J. Algebra **491** (2017), 314–342. MR3699099

[Lam05] T. Y. Lam, *Introduction to quadratic forms over fields*, Graduate Studies in Mathematics, vol. 67, American Mathematical Society, Providence, RI, 2005. MR2104929

[Lan52] Serge Lang, *On quasi algebraic closure*, Ann. of Math. (2) **55** (1952), 373–390. MR0046388

[Lan56] Serge Lang, *Algebraic groups over finite fields*, Amer. J. Math. **78** (1956), 555–563. MR0086367

[Lan64] Serge Lang, *Algebraic numbers*, Addison-Wesley Publishing Co., Inc., Reading, Mass.-Palo Alto-London, 1964. MR0160763

[Len74] H. W. Lenstra Jr., *Rational functions invariant under a finite abelian group*, Invent. Math. **25** (1974), 299–325. MR0347788

[Lie08] Max Lieblich, *Twisted sheaves and the period-index problem*, Compos. Math. **144** (2008), no. 1, 1–31. MR2388554

[Lie17] Christian Liedtke, *Morphisms to Brauer-Severi varieties, with applications to del Pezzo surfaces*, Geometry over nonclosed fields, Simons Symp., Springer, Cham, 2017, pp. 157–196. MR3644253

[Lin40] Carl-Erik Lind, *Untersuchungen über die rationalen Punkte der ebenen kubischen Kurven vom Geschlecht Eins* (German), Thesis, University of Uppsala, 1940. MR0022563

[Mae89] Takashi Maeda, *Noether's problem for A_5*, J. Algebra **125** (1989), no. 2, 418–430. MR1018955

[Man71] Y. I. Manin, *Le groupe de Brauer-Grothendieck en géométrie diophantienne*, Actes du Congrès International des Mathématiciens (Nice, 1970), Gauthier-Villars, Paris, 1971, pp. 401–411. MR0427322

[Man86] Yu. I. Manin, *Cubic forms. Algebra, geometry, arithmetic*, 2nd ed., North-Holland Mathematical Library, vol. 4, North-Holland Publishing Co., Amsterdam, 1986. MR833513

[MH73] John Milnor and Dale Husemoller, *Symmetric bilinear forms*, Ergebnisse der Mathematik und ihrer Grenzgebiete, Band 73, Springer-Verlag, New York-Heidelberg, 1973. MR0506372

[Mic14] Ivo M. Michailov, *Noether's problem for abelian extensions of cyclic p-groups*, Pacific J. Math. **270** (2014), no. 1, 167–189. MR3245853

[Mil80] James S. Milne, *Étale cohomology*, Princeton Mathematical Series, vol. 33, Princeton University Press, Princeton, N.J., 1980. MR559531

[Mor12a] Primož Moravec, *Groups of order p^5 and their unramified Brauer groups*, J. Algebra **372** (2012), 420–427. MR2990018

[Mor12b] Primož Moravec, *Unramified Brauer groups of finite and infinite groups*, Amer. J. Math. **134** (2012), no. 6, 1679–1704. MR2999292

[Mor14] Primož Moravec, *Unramified Brauer groups and isoclinism*, Ars Math. Contemp. **7** (2014), no. 2, 337–340. MR3240441

[MP05] Yuri Ivanovic Manin and Alexei A. Panchishkin, *Introduction to modern number theory. Fundamental problems, ideas and theories*, 2nd ed., Encyclopaedia of Mathematical Sciences, vol. 49, Springer-Verlag, Berlin, 2005. Translated from the Russian. MR2153714

[MS82] A. S. Merkur′ev and A. A. Suslin, *K-cohomology of Severi-Brauer varieties and the norm residue homomorphism* (Russian), Izv. Akad. Nauk SSSR Ser. Mat. **46** (1982), no. 5, 1011–1046, 1135–1136. MR675529

[NS17] J. Nicaise and E. Shinder. *The motivic nearby fiber and degeneration of stable rationality*, arXiv:1708.02790, 2017.

[Oka16a] T. Okada. *Stable rationality of cyclic covers of projective spaces*, arXiv:1604.08417, 2016.

[Oka16b] T. Okada. *Stable rationality of orbifold fano threefold hypersurfaces*, arXiv:1608.01186, 2016.

[Oka17a] T. Okada. *Smooth weighted hypersurfaces that are not stably rational*, arXiv:1709.07748, 2017.

[Oka17b] T. Okada. *Stable rationality of index one Fano hypersurfaces containing a linear space*, arXiv:1709.07757, 2017.

[OVV07] D. Orlov, A. Vishik, and V. Voevodsky, *An exact sequence for $K^M_*/2$ with applications to quadratic forms*, Ann. of Math. (2) **165** (2007), no. 1, 1–13. MR2276765

[Pey93] Emmanuel Peyre, *Unramified cohomology and rationality problems*, Math. Ann. **296** (1993), no. 2, 247–268. MR1219902

[Pey08] Emmanuel Peyre, *Unramified cohomology of degree 3 and Noether's problem*, Invent. Math. **171** (2008), no. 1, 191–225. MR2358059

[Pir16] A. Pirutka. *Varieties that are not stably rational, zero-cycles and unramified cohomology*, arXiv:1603.09261, 2016.

[Pla09] Bernat Plans, *On Noether's problem for central extensions of symmetric and alternating groups*, J. Algebra **321** (2009), no. 12, 3704–3713. MR2517810

[Pla17] Bernat Plans, *On Noether's rationality problem for cyclic groups over* $\mathbb{Q}$, Proc. Amer. Math. Soc. **145** (2017), no. 6, 2407–2409. MR3626499

[Poo10] Bjorn Poonen, *Insufficiency of the Brauer-Manin obstruction applied to étale covers*, Ann. of Math. (2) **171** (2010), no. 3, 2157–2169. MR2680407

[Poo17] B. Poonen. *Rational points on varieties*, Graduate Studies in Mathematics, vol. 186, American Mathematical Society, Providence, RI, 2017.

[Pro10] Yuri G. Prokhorov, *Fields of invariants of finite linear groups*, Cohomological and geometric approaches to rationality problems, Progr. Math., vol. 282, Birkhäuser Boston, Inc., Boston, MA, 2010, pp. 245–273. MR2605172

[Pro15] Yuri Prokhorov, *On stable conjugacy of finite subgroups of the plane Cremona group, II*, Michigan Math. J. **64** (2015), no. 2, 293–318. MR3359027

[PS16] V. Przyjalkowski and A. Shramov, *Double quadrics with large automorphism groups* (Russian), Tr. Mat. Inst. Steklova **294** (2016), no. Sovremennye Problemy Matematiki, Mekhaniki i Matematicheskoĭ Fiziki. II, 167–190. MR3628499

[Rei42] Hans Reichardt, *Einige im Kleinen überall lösbare, im Grossen unlösbare diophantische Gleichungen* (German), J. Reine Angew. Math. **184** (1942), 12–18. MR0009381

[Roq63] Peter Roquette, *On the Galois cohomology of the projective linear group and its applications to the construction of generic splitting fields of algebras*, Math. Ann. **150** (1963), 411–439. MR0154888

[Roq64] Peter Roquette, *Isomorphisms of generic splitting fields of simple algebras*, J. Reine Angew. Math. **214/215** (1964), 207–226. MR0166215

[Ros96] Markus Rost, *Chow groups with coefficients*, Doc. Math. **1** (1996), no. 16, 319–393. MR1418952

[Sal84] David J. Saltman, *Noether's problem over an algebraically closed field*, Invent. Math. **77** (1984), no. 1, 71–84. MR751131

[San81] J.-J. Sansuc, *Groupe de Brauer et arithmétique des groupes algébriques linéaires sur un corps de nombres* (French), J. Reine Angew. Math. **327** (1981), 12–80. MR631309

[SB89] N. I. Shepherd-Barron, *Invariant theory for S_5 and the rationality of M_6*, Compositio Math. **70** (1989), no. 1, 13–25. MR993171

[SB04] Nicholas I. Shepherd-Barron, *Stably rational irrational varieties*, The Fano Conference, Univ. Torino, Turin, 2004, pp. 693–700. MR2112599

[Sch17a] S. Schreieder. *On the rationality problem for quadric bundles*, arXiv:1706.01356, 2017.

[Sch17b] S. Schreieder. *Quadric surface bundles over surfaces and stable rationality*, arXiv:1706.01358, 2017.

[SdJ10] Jason Starr and Johan de Jong, *Almost proper GIT-stacks and discriminant avoidance*, Doc. Math. **15** (2010), 957–972. MR2745688

[Sel51] Ernst S. Selmer, *The Diophantine equation $ax^3 + by^3 + cz^3 = 0$*, Acta Math. **85** (1951), 203–362. MR0041871

[Ser65] Jean-Pierre Serre, *Cohomologie galoisienne* (French), 3rd ed., Lecture Notes in Mathematics, vol. 5, Springer-Verlag, Berlin-New York, 1965. With a contribution by Jean-Louis Verdier. MR0201444

[Ser70] Jean-Pierre Serre, *Cours d'arithmétique* (French), Collection SUP: "Le Mathématicien", vol. 2, Presses Universitaires de France, Paris, 1970. MR0255476

[Ser79] Jean-Pierre Serre, *Local fields*, Graduate Texts in Mathematics, vol. 67, Springer-Verlag, New York-Berlin, 1979. Translated from the French by Marvin Jay Greenberg. MR554237

[SGA71] *Revêtements étales et groupe fondamental (SGA 1)* (French), Documents Mathématiques (Paris) [Mathematical Documents (Paris)], vol. 3, Société Mathématique de France, Paris, 2003. Séminaire de géométrie algébrique du Bois Marie 1960–61. [Algebraic Geometry Seminar of Bois Marie 1960-61]; Directed by A. Grothendieck; With two papers by M. Raynaud; Updated and annotated reprint of the 1971 original [Lecture Notes in Math., 224, Springer, Berlin; MR0354651 (50 #7129)]. MR2017446

[Sha90] I. R. Shafarevich, *The Lüroth problem* (Russian), Trudy Mat. Inst. Steklov. **183** (1990), 199–204, 229. Translated in Proc. Steklov Inst. Math. **1991**, no. 4, 241–246; Galois theory, rings, algebraic groups and their applications (Russian). MR1092032

[Shi17] Evgeny Shinder, *The Bogomolov-Prokhorov invariant of surfaces as equivariant cohomology*, Bull. Korean Math. Soc. **54** (2017), no. 5, 1725–1741. MR3708807

[Sko99] Alexei N. Skorobogatov, *Beyond the Manin obstruction*, Invent. Math. **135** (1999), no. 2, 399–424. MR1666779

[Sko01] Alexei Skorobogatov, *Torsors and rational points*, Cambridge Tracts in Mathematics, vol. 144, Cambridge University Press, Cambridge, 2001. MR1845760

[Sko09] Alexei Skorobogatov, *Descent obstruction is equivalent to étale Brauer-Manin obstruction*, Math. Ann. **344** (2009), no. 3, 501–510. MR2501300

[Sme17] Arne Smeets, *Insufficiency of the étale Brauer-Manin obstruction: towards a simply connected example*, Amer. J. Math. **139** (2017), no. 2, 417–431. MR3636635

[SS91] P. Salberger and A. N. Skorobogatov, *Weak approximation for surfaces defined by two quadratic forms*, Duke Math. J. **63** (1991), no. 2, 517–536. MR1115119

[Sta09] Jason Michael Starr, *Arithmetic over function fields*, Arithmetic geometry, Clay Math. Proc., vol. 8, Amer. Math. Soc., Providence, RI, 2009, pp. 375–418. MR2498066

[Ste77] Robert Steinberg, *On theorems of Lie-Kolchin, Borel, and Lang*, Contributions to algebra (collection of papers dedicated to Ellis Kolchin), Academic Press, New York, 1977, pp. 349–354. MR0466336

[Ste84] Philippe A. J. Steiner, *Groupe de Brauer des corps de fractions rationnelles à coefficients complexes* (French), Enseign. Math. (2) **30** (1984), no. 1-2, 115–140. MR743673

[STVA14] Cecília Salgado, Damiano Testa, and Anthony Várilly-Alvarado, *On the unirationality of del Pezzo surfaces of degree 2*, J. Lond. Math. Soc. (2) **90** (2014), no. 1, 121–139. MR3245139

[Swa69] Richard G. Swan, *Invariant rational functions and a problem of Steenrod*, Invent. Math. **7** (1969), 148–158. MR0244215

[SX17] Jason Michael Starr, *Rational points of rationally simply connected varieties* (English, with English and French summaries), Variétés rationnellement connexes: aspects géométriques et arithmétiques, Panor. Synthèses, vol. 31, Soc. Math. France, Paris, 2010, pp. 155–221. MR2931862

[Tab15] Gonçalo Tabuada, *Weil restriction of noncommutative motives*, J. Algebra **430** (2015), 119–152. MR3323978

[Tig87] Jean-Pierre Tignol, *On the corestriction of central simple algebras*, Math. Z. **194** (1987), no. 2, 267–274. MR876236

[Tot16] Burt Totaro, *Hypersurfaces that are not stably rational*, J. Amer. Math. Soc. **29** (2016), no. 3, 883–891. MR3486175

[Tre91] S. L. Tregub, *Birational equivalence of Brauer-Severi manifolds* (Russian), Uspekhi Mat. Nauk **46** (1991), no. 6(282), 217–218; English transl., Russian Math. Surveys **46** (1991), no. 6, 229. MR1164209

[Tre14] Andrey S. Trepalin, *Rationality of the quotient of $\mathbb{P}^2$ by finite group of automorphisms over arbitrary field of characteristic zero*, Cent. Eur. J. Math. **12** (2014), no. 2, 229–239. MR3130680

[Vis05] Angelo Vistoli, *Grothendieck topologies, fibered categories and descent theory*, Fundamental algebraic geometry, Math. Surveys Monogr., vol. 123, Amer. Math. Soc., Providence, RI, 2005, pp. 1–104. MR2223406

[Voe03] Vladimir Voevodsky, *Motivic cohomology with $\mathbf{Z}/2$-coefficients*, Publ. Math. Inst. Hautes Études Sci. **98** (2003), 59–104. MR2031199

[Voi15] Claire Voisin, *Unirational threefolds with no universal codimension* 2 *cycle*, Invent. Math. **201** (2015), no. 1, 207–237. MR3359052

[Vos67] V. E. Voskresenskiĭ, *On two-dimensional algebraic tori. II* (Russian), Izv. Akad. Nauk SSSR Ser. Mat. **31** (1967), 711–716. MR0214597

[Vos98] V. E. Voskresenskiĭ, *Algebraic groups and their birational invariants*, Translations of Mathematical Monographs, vol. 179, American Mathematical Society, Providence, RI, 1998. Translated from the Russian manuscript by Boris Kunyavski [Boris È. Kunyavskiĭ]. MR1634406

[Wei89] F. W. Weinstein. *On birational automorphisms of Severi–Brauer surfaces*. Prepr. Rep. Math. Univ. Stockholm, 1989.

[Wit16] O. Wittenberg. *Rational points and zero-cycles on rationally connected varieties over number fields*, arXiv:1604.08543, 2016.

[Zag77] A. A. Zagorskiĭ, *Three-dimensional conic bundles* (Russian), Mat. Zametki **21** (1977), no. 6, 745–758. MR0463181

[Zar74] Ju. G. Zarhin, *Noncommutative cohomology and Mumford groups* (Russian), Mat. Zametki **15** (1974), 415–419. MR0354612

[Zho15] Jian Zhou, *Rationality for subgroups of S_6*, Comm. Algebra **43** (2015), no. 7, 2724–2738. MR3354057

Index

Selected Published Titles in This Series

246 **Sergey Gorchinskiy and Constantin Shramov,** Unramified Brauer Group and Its Applications, 2018
245 **Takeshi Saito,** Fermat's Last Theorem, 2014
244 **Atsushi Moriwaki,** Arakelov Geometry, 2014
243 **Takeshi Saito,** Fermat's Last Theorem, 2013
242 **Nobushige Kurokawa, Masato Kurihara, and Takeshi Saito,** Number Theory 3, 2012
241 **O. A. Logachev, A. A. Salnikov, and V. V. Yashchenko,** Boolean Functions in Coding Theory and Cryptography, 2012
240 **Kazuya Kato, Nobushige Kurokawa, and Takeshi Saito,** Number Theory 2, 2011
239 **I. Ya. Novikov, V. Yu. Protasov, and M. A. Skopina,** Wavelet Theory, 2011
238 **Leonid L. Vaksman,** Quantum Bounded Symmetric Domains, 2010
237 **Hitoshi Moriyoshi and Toshikazu Natsume,** Operator Algebras and Geometry, 2008
236 **Anatoly A. Goldberg and Iossif V. Ostrovskii,** Value Distribution of Meromorphic Functions, 2008
235 **Mikio Furuta,** Index Theorem. 1, 2007
234 **G. A. Chechkin, A. L. Piatnitski, and A. S. Shamaev,** Homogenization, 2007
233 **A. Ya. Helemskii,** Lectures and Exercises on Functional Analysis, 2006
232 **O. N. Vasilenko,** Number-Theoretic Algorithms in Cryptography, 2007
231 **Kiyosi Itô,** Essentials of Stochastic Processes, 2006
230 **Akira Kono and Dai Tamaki,** Generalized Cohomology, 2006
229 **Yu. N. Lin′kov,** Lectures in Mathematical Statistics, 2005
228 **D. Zhelobenko,** Principal Structures and Methods of Representation Theory, 2006
227 **Takahiro Kawai and Yoshitsugu Takei,** Algebraic Analysis of Singular Perturbation Theory, 2005
226 **V. M. Manuilov and E. V. Troitsky,** Hilbert C^*-Modules, 2005
225 **S. M. Natanzon,** Moduli of Riemann Surfaces, Real Algebraic Curves, and Their Superanalogs, 2004
224 **Ichiro Shigekawa,** Stochastic Analysis, 2004
223 **Masatoshi Noumi,** Painlevé Equations through Symmetry, 2004
222 **G. G. Magaril-Il'yaev and V. M. Tikhomirov,** Convex Analysis: Theory and Applications, 2003
221 **Katsuei Kenmotsu,** Surfaces with Constant Mean Curvature, 2003
220 **I. M. Gelfand, S. G. Gindikin, and M. I. Graev,** Selected Topics in Integral Geometry, 2003
219 **S. V. Kerov,** Asymptotic Representation Theory of the Symmetric Group and its Applications in Analysis, 2003
218 **Kenji Ueno,** Algebraic Geometry 3, 2003
217 **Masaki Kashiwara,** D-modules and Microlocal Calculus, 2003
216 **G. V. Badalyan,** Quasipower Series and Quasianalytic Classes of Functions, 2002
215 **Tatsuo Kimura,** Introduction to Prehomogeneous Vector Spaces, 2002
214 **L. Š. Grinblat,** Algebras of Sets and Combinatorics, 2002
213 **V. N. Sachkov and V. E. Tarakanov,** Combinatorics of Nonnegative Matrices, 2002
212 **A. V. Mel′nikov, S. N. Volkov, and M. L. Nechaev,** Mathematics of Financial Obligations, 2002
211 **Takeo Ohsawa,** Analysis of Several Complex Variables, 2002

For a complete list of titles in this series, visit the
AMS Bookstore at **www.ams.org/bookstore/mmonoseries/**.